My Big World of Wonder

# My Big World of Wonder

Activities for Learning About Nature
and Using Natural Resources Wisely

*Sherri Griffin*

Redleaf Press
St. Paul, Minnesota
www.redleafpress.org

Published by Redleaf Press
a division of Resources for Child Caring
10 Yorkton Court
St. Paul, MN 55117
Visit us online at www.redleafpress.org

Cover designed by Ryan Huber Scheife.
Interior designed by Jesse Singer and typeset in Vendetta.
Author photograph by Andy Bonderer.
Grateful acknowledgment is made to John Griffin for permission to use his poetry.

Redleaf Press books are available at a special discount when purchased in bulk for
special premiums and sales promotions. For details, contact the sales manager at
800-423-8309.

*Library of Congress Cataloging-in-Publication Data*
Griffin, Sherri.
   My big world of wonder : activities for learning about nature and using natural
resources wisely / Sherri Griffin.
      p. cm.
   ISBN 1-929610-57-2 (pbk.)
   1. Science—Study and teaching (Early childhood)—Activity programs. 2. Nature
conservation—Study and teaching (Early childhood)—Activity programs. 3. Nature
study—Activity programs. I. Title.
   LB1139.5.S35G75 2004
      372.3'5—dc22

                                 2004012609

Manufactured in the United States of America
11 10 09 08 07 06 05 04     1 2 3 4 5 6 7 8

## Dedication

*To the children and families of Millersburg Preschool and Alumni School*
*Thank you for many years of planting and harvesting these seeds*

# Contents

# INTRODUCTION

*The conservation conscience must begin with the young, and there should be opportunity for its blooming.*

(Swift 1967)

CONSERVATION EDUCATION INCLUDES all activities and experiences that result in learning about how people depend on natural resources for needs and wants—how we use and possibly abuse them. It's more than nature study. It involves people's use of *all* natural resources: air, water, minerals, soil, land, and all life forms. The classic textbook definition of conservation is "the wise use of natural resources." Another useful definition comes from the renowned conservation philosopher, Aldo Leopold (1949), who wrote: "Conservation is a state of harmony between men and the land." It is important to note that in order to determine the best use of natural resources, people have to have a basic understanding of nature and the intricacies of how ecosystems operate. Nature study is tied to understanding and practicing conservation.

When I first began teaching, I thought conservation meant saving natural resources. Like many people, I assumed that to conserve a natural resource meant not using it. However, after studying and learning about conservation, I now understand there are three levels of conservation effort: preservation, restoration, and management (Missouri Department of Conservation 1990).

*Preservation,* the first level of conservation, means saving a resource—using little or none of it. For some resources—such as true wilderness, endangered species of plants and animals, small tracts of unique ecosystems, or historically important buildings—this is the only possible method of conservation (Missouri Department of Conservation 1990, 2).

*Restoration* is the second level of conservation. Restoration implies a long-term effort to reestablish the original quality that once existed in the resource being restored. It may mean the return of worn-out farmland to productivity, the restocking of a wildlife species to an area from which it had been depleted, the replanting of denuded forest land, the grading and seeding of barren strip-mined areas, or the reflooding of a drained waterfowl marsh.

*Management* is the third level of conservation, which requires that people make decisions and implement practices. People often don't realize that the decisions they

> The term "wise use" as employed by this author does not refer to any political entity or use by groups with a hidden agenda. This historic term means exactly what is implied and carries its message not only in its straightforward meaning but also through the philosophies of such early conservation thinkers as Aldo Leopold and Teddy Roosevelt.

make about their backyards become land management decisions that have an impact on whole ecosystems. For example, I live in a small, rural community on a gravel county road. Most of my neighbors live on five or more acres of land. Some have chosen to clear their acreage and plant and mow grass; some harvest hay, plant gardens, or use the land for livestock; some plant warm-season grasses and manage controlled burns on a regular basis; and still others let the land proceed as it would naturally. Each of these individuals, whether consciously or not, is making a decision about how to manage a natural resource, thus affecting the wildlife sharing the habitat.

As I write this book, devastating mudslides in California are dominating the news. Fall wildfires killed the vegetation that held the soil in place. Winter rains are washing away the soil, taking homes with it. In order to better understand the interdependency between these events and how they may have been managed differently, it's important to understand the ecosystem of the area. Vegetation holds the soil in place. Within this wildland ecosystem, fires happen naturally and are vital to the health of the ecosystem.

Fire spawns a period of rebirth and vigor by removing dead and diseased vegetation while releasing nutrients locked in mature plants and organic litter. Fires are necessary to regenerate the ecosystem. When people decided to use the land to build homes and businesses, they changed the naturally occurring ecosystem. They managed the land by suppressing fires but little was done to prune the old, dead, and diseased vegetation resulting in a buildup of highly flammable materials. When fires finally occurred, they were large and uncontrollable. Dead and healthy vegetation alike were destroyed, leaving nothing to hold the soil in place. People lost homes in the fires as well as in the resulting mudslides. People decided how to manage the natural resource with little understanding of the delicate balance maintained in the ecosystem and the necessity of fire in the healthy cycle.

While much of this information seems beyond the understanding of young children, the attitudes they develop about the earth's natural resources begin at an early age. Conservation is a philosophy of daily living that reflects a pattern of people's behavior with respect to our life-sustaining environment. This philosophy extends to everyday decisions in the classroom. For example, when serving snack, how is the food distributed? When individuals serve themselves, someone invariably takes more than she can eat. Someone else at the table may not have enough to satisfy his hunger. When an adult serves out the portions, again some children have more than they can eat while others want more. What is the fair way to distribute snack where children take responsibility? Discussions about allocation of snack resources result in a discussion about conservation or "wise use."

Another example from my classroom concerned one of the literacy experiences that children especially enjoy: character suitcases. Children sign up to take these suitcases home on a rotating basis. This one was based upon the book *Nicky the Nature Detective*. Inside the suitcase was the book, a Nicky doll, binoculars, a compass, insect box, magnifying glass, blank book, and colored pencils. One child had taken Nicky home and kept forgetting to bring her back. After several weeks, the family wrote me a note saying they could no longer find all of the parts. A classroom discussion followed about how to handle the suitcase issue. I was amazed to find that all three levels of conservation were involved in this discussion. One child thought the suitcases shouldn't go home with anyone any more, they should remain at school for children to use (preservation). Another thought the family should have to buy all new materials for the suitcase (restoration). Still another child thought the family should not get any more suitcases until they returned the lost materials (management). Although this discussion wasn't about natural resources or a topic considered relevant to conservation, it was about a classroom resource and was conservation minded. The way problems are handled in the classroom gives children strategies for handling problems in everyday life. Conservation ethic is taught whether teachers intend it or not.

## What Can Children Learn from Conservation and Nature Activities?

[A] way of looking after our world, just as we would look after our homes. Use our intelligence and common sense to keep it safe, to make it comfortable and beautiful, to repair any damage and upset, to preserve it for the future
(Ingpen and Dunkle 1987, 6)

Young children are developing lifelong patterns of living and attitudes about the earth and its natural resources. Noticing changes in their environment and discussing how those changes have an impact on the other natural resources is an important part of this. Talk about the vacant lot down the street being renovated as a park, the field down the road being bulldozed to create a shopping mall, or the cleanup of the local creek—all are land-use decisions that have an impact on all of the natural resources in the area.

This past fall, my preschoolers discovered a number of holes in our play yard. Each hole had a small mound of soil next to it. There was much discussion and speculation as to what was creating these holes—snakes, moles, worms, rabbits, other children, and so on. A popular activity was to stomp the dirt mounds or scrape the soil back into the holes.

As we were coming inside one day, the children discovered a crayfish in the parking lot. They immediately wanted to bring it into the classroom. An appropriate container was sought and the crayfish joined us at group time. The children gathered sketchboards, pencils, magnifying glasses, flashlights, and a Missouri field guide on crayfish to aid them in their explorations. We looked *crayfish* up in our field guide and discovered they are crustaceans and are also called *crawdads*. While some species live in ponds or streams, others live in burrows that can be identified by their above-ground "chimneys" made of mud balls from digging. These chimneys were the soil mounds next to the holes on our play yard.

Upon further examination, the children discovered that this crayfish was a she. This mother crayfish was carrying her young under her tail. They were hanging on to the small appendages (*swimmerets*) on her abdomen. The children speculated as to why she wasn't called a "crawmom." They then decided to sketch this unusual mother. Much time was spent noticing and counting legs, looking at body parts, and discussing her antennae. Sketch boards helped make the drawing easier. Flashlights and magnifying glasses assisted the children in making more detailed observations. After discussing, examining, and sketching the "crawmom," the children decided to release her back on the playground. We watched as she maneuvered her way off the stump and into the grass to find her home.

From this simple and spontaneous activity, the children gained a new understanding of conservation. For example, they learned facts about the earth—in this particular case it was information about the crayfish. However, this might as well include information about the child's own corner of the world or issues about the planet as a whole. Children are intent on discovering the world around them—they can gather information and facts about the interdependency of people and natural resources when we capitalize on this "discovery learning" and support explorations with resources and futher experiences.

The crayfish helped the children develop *respect for nature and all living things*. Young children are egocentric and only see the world from their own perspective. Conservation and nature activities make the most of that focus while helping to broaden their horizons. When the children considered what would happen to the "crawmom" if it wasn't released in an appropriate habitat, they moved beyond their own wants while still honoring the need to study and develop an understanding of what a crayfish is, where it lives, how it moves, and what it eats. This exploration not only gave them appreciation for the mother crayfish but changed their behavior in the play yard. Now they actively looked for more signs of the special inhabitants who shared their play area.

Conservation education ultimately enables children to *better care for the earth and its inhabitants*. The children's understanding of the crayfish's habitat helped them care for their own environment in a different way. They no longer disturbed the mounds of dirt next to the holes. In fact, they often went to great lengths to avoid stepping on them when running and playing.

The children also gained *an appreciation of the beauty of the world*. Sketching the crayfish allowed them to consider her unique features and how beautifully she was suited for her environment. Conservation and nature activities will help children *be more observant and aware of their surroundings* as well as of the earth in general.

The crayfish experience encouraged these children *to explore, question, and experiment to find answers for themselves.* As a result of this experience, they became crayfish experts. They also learned about the study of nature. Sketching, observing, and seeking answers in resources provided them with the opportunity to use tools to support their understanding. The whole experience, which lasted less than an hour, changed the way they viewed and cared for their play environment and the inhabitants in it.

## How Can Teachers Help?

*If a child is to keep alive his inborn sense of wonder . . . he needs the companionship of at least one adult who can share it, rediscovering with him the joy, excitement and mystery of the world we live in.*

(Carson 1956, 45)

Activities and experiences should *emphasize feelings rather than knowledge* for three- to five-year-old children. The seeds of awareness, of feeling and caring, will grow into knowledge and wisdom if sowed within the young child. Rachel Carson (1956) called it the "sense of wonder" and wrote a book of the same title for parents and teachers of young children. Her immortal message encouraged teachers and parents to make discovery fun and to stimulate children's natural sense of wonder. Later, when children are ready, they will assimilate the facts that correspond to personal feelings.

There are many ways that teachers can assist children in learning to use natural resources wisely. Undoubtedly, one of the most important things to consider is to *use developmentally appropriate, hands-on, minds-on learning experiences.* Whenever possible, capitalize on the child's natural curiosities. Encourage children to use each of their senses. Young children, like adults, learn best when they are actively involved. Early childhood educators must take advantage of what we know about how young children grow and learn in developing experiences.

Playing to children's inborn sense of wonder, rather than concentrating on knowledge acquisition, is important. Many adults have lost that sense of wonder or are afraid of exploring or studying something we know little about. *Don't be afraid to explore unfamiliar phenomena.* It's important that adults have opportunities to keep that inborn sense of wonder alive. Some of the most interesting and successful studies I have had with young children were when I was able to learn along with them. I will never forget our discovery of the small crayfish under the mother's tail; the realization that ticks have eight legs and are arachnids rather than insects; or our observation of the egg sack being dragged behind a wolf spider that soon hatched hundreds of baby spiders, which she carried on her back. These were all discoveries we made together, and I'm certain my delight and wonder were as evident to the children as their's were to me.

- *Take advantage of teachable moments.* These are times when a child or group of children have expressed an interest in something. Seize this opportunity to expand upon the expressed interest—it may not be there later when you are ready to do a unit on it. A study on crayfish that I initiated wouldn't have been nearly as successful or as valuable as what occurred by taking advantage of the teachable and learnable moment. All individuals learn best when they are interested in the topic being explored. Following the children's lead promotes learning experiences where everyone is actively involved in the learning, including you as teacher. Think of the opportunity that would have been missed if we hadn't invited the "crawmom" into our classroom!

Early childhood educators practice conservation on a daily basis. For example, the recycling that occurs with art materials alone often goes unnoticed. As teachers, we should *encourage conservation or wise use of all materials in and outside the classroom.* Set up recycling in the classroom and provide bins for sorting and recycling non-reuseable items. Point out the reuse of found materials that occurs in the classroom on a daily basis. Children follow and learn from adult models.

- *Provide opportunities for children to explore and discover explanations for themselves.* When children ask questions, provide experiences that will allow them to discover answers for themselves. Often, errors in children's thinking can also lead toward discovery learning. When my husband and I first began building our house, our five-year-old nephew was staying with us. At that time our dryer was not vented, which resulted in a number of cobwebs on the insulation near the dryer. I noticed him walking widely around this area as if avoiding it. I asked him why, and he quickly informed me that he was afraid of the spiders in the spiderwebs. In my all-knowing, grown-up mind, I solved this problem by simply telling him that those were cobwebs rather than spiderwebs. The next day, I found him with his magnifying glass and science journal carefully looking at the cobwebs. Again, I asked what he was doing. He quickly informed me he was looking for cobs. At this point I stepped back and allowed the study to pursue a more natural course. I had contributed to his error in thinking and I needed to step back and allow him to pursue this course of investigation for himself. If exploration is not possible, *answer questions simply and honestly,* being careful to only provide as much information as the child requests and making sure that the explanation you provide makes sense on a child's level.

- *Encourage children to be observant.* Provide experiences where children can explore with all their senses. Children learn by your example—they learn to observe, sketch, and record data about phenomena. If you don't consider yourself an artist and scientist, the children won't think that of themselves either. Ask direct questions that will focus and challenge children's thinking. Bring the observation to the child's developmental level. For example, encourage discussion about color, size, shape, texture, or smell rather than providing factual information. During their study of crayfish, children were interested in counting the number of legs on the crayfish as well as her color and the size of her young. These were all important observations for their learning.

- *Use a variety of media and provide a variety of experiences.* Refrain from providing isolated experiences, but rather integrate as many of the young child's activities as possible. One early spring, my preschoolers and I took a field trip to a local conservation area to look for signs of spring. One five-year-old was leading some of our new three-year-olds through the area providing information about the trees. I heard him tell the threes that soon the leaves (the ground was covered with leaf litter from the previous fall) would be "magicked" back up onto the trees. The three-year-olds looked at him with wide, believing, amazed eyes. I knew that no amount of explanation from me could refute what this worldly five-year-old had shared. After contemplating the situation, I planned a series of field trips to the area at approximately two-week intervals. Each time we visited, the children were equipped with their magnifying glasses and science journals. As we spent that spring exploring the trees, buds, and leaf litter, the children soon discovered how the leaf litter decayed to form soil and the new leaves came from the buds on the trees. Repeating this field trip provided the children with the opportunity to change their ideas about leaves being "magicked" back onto the trees with little explanation on my part.

- *Provide examples of good and poor use of resources.* Children need to see that natural resources can be used in both productive and nonproductive ways. This might be as simple as talking about erosion on the play yard or as complex as visiting a landfill. Set a good example in your own treatment of the natural resources of the earth. Remember that your actions speak much louder than your words and children will follow your lead.

- *Take advantage of the resources offered in your state and local area.* State governments generally have a department of conservation, department of natural resources, wildlife and fisheries division, or some such agency. These state agencies offer resources and materials for teachers who want to find local and regional information. In addition, they often have specific personnel to deal with various topics surrounding conservation and wildlife. Most every area in the country has some type of wildlife officer. These game wardens or conservation agents are often assigned to a specific county. If you don't know who your local conservation agent is, a good place to find that information is the county sheriff's office.

Inviting an expert into the classroom provides children with a new experience—even with a familiar topic. For example, a conservation agent recently visited the preschool and talked with the children about turtles. He brought along several turtles to help them gain a better understanding of turtles and their habitat. The children counted each turtle's toes to help identify the type of turtle. They also counted the rings on the plates of the turtles' shells to age them. The agent provided the children with a pamphlet and encouraged them to find and read about the turtles they saw. Discussing ways that turtles help people, and the interdependence of people and turtles on the environment, addressed several conservation concepts. Habitat was addressed when the children discussed where turtles live and what they eat. Sketching the turtles was a key part of the experience for the children. They learned about researching a topic by reading numerous books about turtles and searching for information after the presentation. Most of the children went home and talked about the turtle visit with their families, recalling what they learned. Ever since that experience, many stories have been told at school about turtles the children had seen and identified during family outings. This one, brief study challenged the children in many different areas and stretched their understanding not only of traditional school topics but about turtles as well. It was an authentic learning situation that was important and real for them, rather than merely something they read from a book.

## Issues Surrounding Conservation Education

As you delve into conservation education with young children, you will most likely come across several issues or controversies. The first of these is *anthropomorphism,* or giving human characteristics to animals. These characteristics include animals talking, expressing complex emotions, wearing clothing, living in homes with furniture as people do, and so on. Much of children's media embraces this concept. The egocentric nature of young children as well as their emerging ability to distinguish reality from fantasy feeds this notion. As a result, many children grow up believing that animals think and respond the same as people. Although some of the learning experiences presented in this book encourage children to act or pretend to be animals, this is done to help them gain understanding of the natural resource. This doesn't mean teachers need to throw out all the anthropomorphic literature or other materials used to help children learn, but rather to discuss the reality versus fantasy nature of anything we share with children.

Another difficult topic for teachers of young children to broach is *hunting.* Hunting is included in *My Big World of Wonder* because animal harvest (hunting, fishing, and trapping) is a management tool for wildlife. *Animal harvest* is the term used in most of the conservation literature. Several stories feature hunters as safe, ethical, and responsible. While it is not a prominent piece of this book, it is part of the underlying conservation ethic. Many species in the wildlife population owe its health to hunters. In many states, hunters were the first to start and fund state wildlife organizations. For example, in 1937, the Missouri white-tailed deer population was nearly nonexistant. As a result, the state officially established the Conservation Commission, which was funded through sales of hunting and fishing permits. The commission began a comprehensive restoration program that included changes in hunting regulations, stronger enforcement, research, livetrapping and distribution, and an educational effort. As a result of those efforts, Missouri now has a healthy deer herd that supports an annual harvest many times larger than the total deer population before the restoration began. However, since the deer herd has no natural predators in the state, without the annual harvest, the deer population would quickly strip the food supply thus causing grave repercussions to its own survival as well as many other natural resources, including people (Missouri Department of Conservation 1990).

Inevitably, along with the discussion of hunters comes the issue of *guns in the classroom.* For many years, I outlawed gun play in my classroom. Although I was married to a law enforcement officer, I felt nothing constructive could come from children pretending to shoot guns. What I found happening in the classroom was that children continued to make and play with guns but became very adept at changing them to something else when I was in the area. Consequently, I missed many of the details of the play and children were not given the opportunity or support to safely explore more appropriate uses of guns.

One day I was observing two boys from across the room. One boy moved behind the sink in our pretend play area with a duck puppet, while the other stood with a long stick-like creation that he had made from Duplo blocks. I was watching to stop the play if it turned out to be gun play. As I observed the boys, the one behind the sink threw the puppet up into the air while the other took aim and shot his play gun. The duck fell and the boy behind the sink picked up the puppet in his mouth and brought it to the shooter. They then proceeded to the play stove and cooked the duck. I wasn't quite sure how to respond.

Gun play was happening in my classroom, but what I observed was really quality play. The children had a purpose, assigned roles, and had to coordinate their efforts. They were exploring what it was like to hunt, retrieve, cook, and eat duck. How could this be bad? Our classroom is a community of learners and all constructive play should be valued. How could I make the decision about this play being inappropriate?

As a result of that experience, I have rethought my policy for gun play. I still do not allow children to bring toy guns into the classroom. However, when guns are created in the classroom, we generally have a class meeting to discuss what the children think might be the most appropriate kinds of things to do about this play so people are not hurt. Generally they decide that guns should not be shot at people, good guys or bad guys. An opportunity is presented to discuss our classroom being safe for everyone and that guns should not be used to hurt people. As a result, children generally decide to use the guns they create for target practice or hunting purposes.

*Fishing* might also be a controversial issue for some teachers. Again, responsibly harvesting fish, whether from a river, lake, or the ocean can promote a healthy population. However, overfishing our oceans has caused depletion of some species to the point of endangerment. A conservation ethic promotes solid reasoning for the management of each species. People ultimately make decisions about what is the wisest use of this natural resource. Exposing children to the idea of fishing and managing this resource on the small scale of their classroom will someday have an impact on how they manage this resource globally.

In many regions of the country, the *harvest of trees* is also a controversial and often highly criticized practice. Nearly every single individual uses some form of forest products on a regular basis. Although it is easy to criticize how this particular resource is managed, it is important to note that it is a renewable resource and as such, the forest (resource) can benefit people and nature. No-use at one extreme and unrestricted logging at the other are not wise choices for our forest lands. Selective, managed harvest will insure the health of the forest, now and in the future. Consequently, it is vital that children be exposed to trees, their ecosystems, and how people use them so they will be able to make future choices concerning the harvest and management of this valuable resource.

Rural versus urban understanding of conservation ethic may also become an issue in the classroom. Because children who live in large cities do not have the same experiences as their peers living in rural areas, teachers may believe that the conservation concepts should be handled differently in the two places. And, in fact, some aspects of the curriculum may change from the city to the country. In order to make the curriculum work for your students, you may want to find out, for example, how many of your students' families hunt or fish, or what kinds of experiences they have had with guns. However, people make wildlife management decisions every day in cities as well as rural areas. The raccoon in the garbage, deer eating flowers, geese eliminating in picnic areas, or trees growing where people want to build homes— solving all of these problems involves making conservation management decisions. It is important to help children gain an understanding of the many conservation management decisions made wherever they live.

Sometimes it is difficult to determine how to handle the discussion of these many management decisions in the classroom. It is important to value individual family background and beliefs while honoring various practices, such as hunting or logging. Ultimately, teachers have to decide where this balance lies in their own classrooms. However these issues are dealt with (or not dealt with) in the classroom, it is important to remember that conservation is a philosophy of daily living. As teachers, we have the power to influence that philosophy whether we want it or not. Only knowledgeable, well-informed people can make decisions about best practices. As early childhood professionals, our responsibilities lie with helping children develop attitudes and gather information about the interdependence of natural resources and people. Conservation-ethic decisions will belong to them in the future.

As a teacher, I have grown and matured in handling these issues throughout the years. When I first began thinking about conservation in my classroom, I considered it just another topic or study that we would engage in on an occasional basis. However, as time has gone by and I have read, studied, and lived the topic, I have come to realize that conservation ethic is addressed in life on a daily basis. It is not something that I decide to teach or not to teach. Conservation ethic is evident in every decision made by me, the children, or our classroom community. My ultimate goal for the children is not that they grow up to be preservationists or hunters but rather that they consider all sides and make decisions based upon a real understanding of ecosystems and how they work. Only then will they be able to handle the "wise use" of our world's natural resources.

Wondering about nature and conservation with the children over the years has also taught me a great deal about teaching and learning. I am a teacher because I love learning. The children are my guides and teachers. Listen, learn, and live with the children and they will guide you to places you never dreamed. Ultimately, I would change Rachel Carson's quote to say, "If a *grown-up* is to keep alive *his/her* inborn sense of wonder . . . *he/she* needs the companionship of at least one *child* who can share it, *discovering* with *him/her* the joy, excitement, and mystery of the world we live in."

## Key Tools for Conservation Study

*It is more important to pave the way for the child to want to know than to put him on a diet of facts he is not ready to assimilate.*

(Carson 1956)

### Science Journals

Science journals are a wonderful tool to help young children be observant and record their ideas and observations while they give you insight into their thinking. Journals can work as an assessment tool, providing valuable information about where a study should next proceed or experiences you need to help children clarify understanding.

The journal can be several pieces of paper stapled together, a small spiral-bound notebook, or something more elaborate. I generally make a science journal for each child at the beginning of the year (see instructions on page 13). Journals can be introduced in a variety of ways. I typically gather children together and focus them on a particular experiment in the classroom or introduce journals during a walk. Another idea is to place them in the science area near an experiment where the children can use them on their own. However, I have found that with this method, not all children will use their journals, and many do not use them in a purposeful manner. Whatever strategy you use to introduce the journals, there are several important points to keep in mind:

- *Model scientific behavior for the children.* This is not so they will copy your writing and thoughts but rather so they will see how important recording data and keeping track of it is. When you as a teacher participate in an activity, children see this as an important, grown-up thing to do. I often talk aloud while writing and drawing in my journal so children can hear my thinking. My journal entries might be a drawing, chart, recipe, or merely a prediction. Just as the children aren't always correct in their thinking, neither am I! My science journal is also where I keep track of my data, thoughts, and ideas. If I place value on my journal, take care of it, and write things that have meaning for me, so will the children.

- *Accept whatever they choose to draw or write in their journals.* Children may be hesitant about writing and recording their thoughts; assure them it is all right to pretend to write or draw pictures of their discoveries. Your expectations are important in how children feel about their entries. Everything—including drawings, scribbles, strings of letters, and invented spelling—should be accepted and acknowledged as writing.

- *Provide science journals for children to use throughout the day.* If you want children to learn to record their data, hypotheses, and conclusions, they need to have access to these journals whenever they feel they have something worth recording. Remember that you are not the only one who can lead an experiment. Once children learn to use their science journals as a tool, they frequently record information and data on their own. Having science journals always available allows you to take advantage of the teachable moment from across the room.

Sketchboards are another tool for recording ideas and observations. I use small, inexpensive clipboards, purchased at our local discount store. However, a sketchboard could be as simple as a piece of sturdy cardboard and paper attached with a rubber band. I use sketchboards with the children when I want to use their work in documentation or display. With sketchboards, I usually provide colored pencils, and the focus is primarily on sketching or drawing.

## Nature Walks

Nature walks are an easy, inexpensive, and readily accessible way to share nature with young children. Don't be reluctant to take children on nature walks because you think you don't know enough about nature. Your goal should be to share the wonder and beauty of the natural world rather than to feed children "a diet of facts" they are not likely to remember. Consciously use all the senses to help young children develop relationships between past experiences and new events. The following list of activities will help you guide children in developmentally appropriate ways while encouraging the development of an awareness of the natural world.

### SCAVENGER HUNTS

Scavenger hunts provide an excellent way to encourage children to stop and look at the environment. Try a different focus with each of the following experiences:

- *Color.* Pass out paint samples (obtained from your local hardware or paint store) before beginning the walk. Challenge children to find natural items that are the

## Journal Instructions

1. Take a large piece of specially decorated paper (at least 8½ by 11 inches) and lay it face down on the table. I usually use paper from the children's art projects.

2. Glue two pieces of mat board (4½ by 6 inches) in the middle of the paper, leaving about ¼ to ½ inch between the two pieces of mat board and about ½ to 1 inch all the way around the outside. Smooth out the paper. (Mat board is available at most frame shops who will often donate their scraps.)

3. Cut away the corners of the paper to the corner of the mat board.

4. Wrap the paper around the mat board, similar to wrapping a present, and glue.

5. Take ten to twelve half-sheets of paper. Fold each in half. Open the paper and place one on top of the other, like a tent. These can be sewn on a sewing machine using a wide stitch (then skip to step 9) or sewn by hand.

6. Take a "yarn darner" needle (or an awl) and poke three holes in the crease of the papers— near the top, bottom, and in the middle.

7. Cut a yard of crochet cotton and wax with beeswax to keep it from cutting the paper.

8. Thread the needle and sew up the pages like a figure 8 (from the side), starting and ending at the top.

9. Glue the first page to one piece of mat board and glue the last page to the other. Smooth out and try opening and shutting the book several times to make sure it stays in place.

same color as their paint chips. Younger children can be challenged to find items the same general color as the paint chip. The experience can be made more challenging for older children by asking them to find items the exact shade of their paint chip. For example, pass out shades of green and challenge children to find the exact shade as their paint chip.

- *Pictures.* Pass out pictures of plants, animals, and other natural items that children might encounter on their walk. Challenge children to find the item on their picture. This works especially well with young children, yet it can be challenging for older children when the items are difficult to find.

- *Texture.* Assign each child a texture to explore. For example, one child might look for smooth items, while others might look for items that are prickly, hard, hairy, soft, sticky, and so on. Older children can feel for opposites in texture.

- *Contrast.* Challenge children to look for contrasts on the nature walk. For example, the driest and wettest place, the coldest and hottest, the place that receives the least and most sunlight, or the oldest and youngest thing.

- *Size.* Challenge children to find items on the walk in relation to size. For example, a plant as tall as your waist, a leaf as wide as your foot, a plant as long as your little finger, or a flower as big as your fist.

- *Sounds.* Play an audio recording of birdcalls for birds whose habitats are at the site of the field trip. While outside, try to locate and identify those birds.

- *Shapes.* Assign children a different shape and then challenge them to find as many items as possible that share the same shape.

- *Animal signs.* Challenge children to look for signs of animals. For example, a place where an animal has eaten, tracks left by an animal, or a nest in a tree.

## EXPLORE AN AREA

This type of experience is different from a scavenger hunt in that children are challenged to focus on a small piece of their environment rather than walking or moving through the entire environment. As with all of these experiences, it is important that you join in the activity and model behavior for the children.

- *Listen to the grass grow.* Ask children to lie on the ground and pretend to be part of the earth. Be sure everyone lays where they are not touching someone else. If children will allow you, have them close their eyes and lightly cover their faces and bodies with dry grass or twigs so they actually look like the ground. Explain that they must be quiet and blend in with their surroundings if they want to hear and see things. After several minutes, begin a discussion of how they felt and what they heard, saw, and smelled.

- *Make a circle of exploring.* Place circles of yarn around a small area in a park, wooded area, or play yard. Provide each child with a magnifying glass and a science journal. Challenge children to find all of the living and nonliving things in their circle of ground. Try this activity in several different habitats and compare the results.

- *Explore a tree trunk.* Challenge children to explore a rotting log or tree trunk in the woods. Again, provide magnifying glasses and science journals for sketching and recording their finds.

- *Explore a habitat.* Select a specific type of habitat for children to explore (such as a fence row, wetland, creek bank, and so on). Return during the different seasons of the year to see how the habitat has changed. Magnifying glasses and science journals will help children focus and record their findings.

## COLLECTING ARTIFACTS

Collecting artifacts and taking them back to the classroom is a wonderful way to extend the nature walk experience. However, there are some things to remember when considering collecting as part of the walk:

- *Check the rules of the area.* Many federal and state department areas are protected, which means that items may not be taken from the area. Be sure to look for posted rules and read them with the children. Model responsibility.

- *Discuss safe items to collect.* Children are often zealous in their collection process—they neglect to consider safety. Be sure to discuss what the children are collecting and how to make sure it is something safe before beginning the walk.

However, it is important to keep these suggestions short and simple as children tend to forget too many instructions.

- *Respect the area.* If the area provides opportunity for collecting, caution the children about only taking what they need for your specific purpose. The rest should be left behind for other users of the habitat to enjoy. Discuss "wise use" of the natural resource and the impact the children might have on the area. For example, when children collect seeds or nuts, collecting all of the acorns they can find might deprive the local wildlife of a food source; or when collecting flowers, picking everything in sight ruins the beauty for the next visitors.

- *Consider alternative methods of collection.* Collecting artifacts from an area doesn't always mean physically removing the item. Photographs and sketches are always possibilities, but audio recordings of sounds in the area can also be a valuable reminder of the experience for children. Listening to these tapes back in the classroom allows children (and you) to notice details missed while in the area. Although more challenging, smells can also be brought back to the classroom. For example, on a walk that my children frequently take, there is a grove of cedars that produces a particularly pungent smell. A small sprig, slightly crushed, is all it takes to instantly transport us back to the area.

- *Always leave the area cleaner than you found it.* Take a trash bag to pick up trash even when this isn't your goal for the walk. You are modeling responsibility and caring for the environment.

## SAFETY TIPS

- *Everyone should be clothed in appropriate apparel.* Socks, comfortable shoes, and long pants will help to avoid problems after the walk begins.

- *Check out the area prior to the walk.* Suit the length of the walk to children and avoid hazardous areas. Look for restroom facilities and determine whether you need to bring water for the hike. Familiarity with an area will help you to create a better pace for the children and avoid accidents.

- *Carry a backpack with emergency information and a first-aid kit.* You might also want to include water, a cell phone, and a light snack. Optional items might include magnifying glasses, science journals, colored pencils, flashlights, a tape recorder, sketchboards, and a camera.

- *Encourage children to look with their eyes rather than disturbing areas.* After they have been examined, remember to return leaf litter and rotting logs to where they were found.

- *Discuss what to do if you encounter a snake.* Children should freeze, then slowly back away. Never attack or attempt to capture the snake. If a child is bitten, the Red Cross advises that the bitten area be kept at or below the level of the heart. Seek medical attention immediately.

- *Learn to recognize poison ivy and poison oak.* I keep a set of leaves, which I have laminated, to help the children and me remember exactly what these plants look like. Poison ivy and poison oak look a great deal alike. The foliage always turns a brilliant red in the fall, enticing young hands to pick it during collecting trips. Find the plant at the beginning of the experience and show it to the children. Repeat the identification often. I teach my children the following lyrics to the tune of "Yankee Doodle" and we sing it periodically throughout the walk:

> *Poison ivy (oak) has three leaves.*
> *White berries grow upon it.*
> *It is food for birds and deer but people should not get near.*
> *Poison ivy (oak) leaves of three.*
> *Poison ivy (oak) let it be.*
> *Bush or vine do not touch it—*
> *Unless you want to itch, itch, itch!*

The song helps children remember what to look for but also encourages them to see the purpose poison ivy or poison oak serves in nature. Use the words most appropriate to the species of poison ivy or poison oak indigenous to your area.

## Organization of the Book

*My Big World of Wonder* is a comprehensive early childhood conservation education program. The program is designed to enable teachers of young children to heighten children's awareness of nature and conservation. The learning events in this book have been designed and prepared with developmentally appropriate practice for three- through five-year-olds in mind. They capitalize on the ways young children learn best—building relationships through active, concrete experiences; manipulating materials; discussing events as they occur; adult modeling; and play. You are the expert on the children in your classroom. Choose activities carefully according to the interest, experience, and developmental understanding of the children involved.

A seasonal framework has been chosen for this book because this approach is generally meaningful to young children and provides a planning model for you. The seasonal format allows you to consider the teachable moments that are likely to occur during a particular time of year. In addition, you can more easily set up a particular activity or learning experience when the phenomena is most easily observed. For example, make track puzzles available during the snowy or rainy season when children are most likely to see the tracks—this encourages children to be observant and ultimately provides more teachable moments.

The seasonal framework also makes sense because all nature study and conservation efforts are tied to the seasons. Hunting, trapping, and fishing regulations are all linked to seasons of the year. For example, many states have both a fall and a spring turkey season. However, hens are often not allowed to be harvested in the spring because they are nesting. To help you expand and develop a study or project around a particular topic, thematic lesson plans are provided in appendix C.

Each season, beginning with fall, is introduced with a general description of the season's environmental changes as well as changes in interactions between people, the environment, and seasonal land-use activities. Not all regions of the country experience the same seasonal changes, but it is important to note that all parts of the country experience some kind of seasonal changes. Tailoring the book and activities to your region of the country will be important to maximize the impact the experiences have in your classroom.

Each activity includes the following:

- Suggestions for the area or learning center where the event might be most successful—art, blocks, bulletin board, field trip, group, large motor, manipulative, music, nutrition, outside, pretend play, science, story, writing, and woodworking

- Background for the teacher as well as a brief description or purpose of what children will gain from the experience

- Materials listed for each activity

- Step-by-step directions for conducting the activity, as well as open-ended questions for discussion starters or assessment

- Suggestions for additional activities to extend or expand the learning event. If the activity calls for flannelboard characters or puzzle pieces, the patterns have been included for your convenience

Appendix A includes lists of children's literature relevant to the thematic topics and appendix B is a reference guide of teacher resources and organizations. Appendix C contains the thematic lesson plans.

The activities and experiences in *My Big World of Wonder* are designed to help you, as a teacher or significant adult in children's lives, connect conservation to everyday living. These are seeds of ideas intended for planting for the future of our environment. Whether you use a few or all of them, you will impact the attitudes and actions of future generations and how they care for the earth.

# References

Carson, Rachel. 1956. *The sense of wonder.* New York: Harper & Row.

Ingpen, Robert, and Margaret Dunkle. 1987. *Conservation: A thoughtful way of explaining conservation to children.* Victoria, Australia: Hill of Content.

Leopold, Aldo. 1949. *A Sand County almanac: And sketches here and there.* New York: Oxford University Press, Inc.

Missouri Department of Conservation. 1990. *An ecological approach to conservation education.* Jefferson City, MO: Conservation Commission of the State of Missouri.

Swift, Ernest. 1967. *The Conservation Saga.* Washington, DC: National Wildlife Federation.

# Fall

CHILDREN ARE ALWAYS EAGER about the approach of fall as seasonal changes trigger changes in plant, animal, and human activity. Yellow school buses appear, friendships are renewed, and a new learning season begins.

Days grow shorter as the sun continues its southerly movement and people take to the outdoors. National Hunting and Fishing Day celebrations are held throughout the nation. All across the United States, fall hunting seasons, carefully monitored by federal and state wildlife agencies, reduce the number of animals that compete for winter food. Elk, deer, moose, and other large mammals begin their ancient courtship rituals all over the country. Their movement during the breeding season makes these large mammals easier to spot but a danger to motorists.

Preparations for winter are in full swing. Crops are harvested, some birds start their seasonal migrations south, many plants, people, and animals begin to prepare their homes and store food for the winter months ahead.

Deciduous trees become a blaze of orange, red, and gold as cooler temperatures and fewer daylight hours cause chemical changes in their leaves. In northern regions of the country, this lasts for several weeks, while the warmer climates experience more subtle color changes for a much shorter period of time. Later in the fall, severe frosts or strong breezes carry the flashy leaves to the ground where they eventually decompose.

On the East Coast of the United States, hurricane season is in full swing, ending in late autumn. In the West, wildfire season is raging but will draw to a close with the start of the rainy season.

The learning experiences in the fall section of *My Big World of Wonder* are designed to capitalize on these naturally occurring phenomena. Topics include the following:

- Trees and leaves
- Seasonal changes
- Harvest—both plant and animal
- Seeds
- Food preservation
- Land use

# What Can You See? ✎ Field Trip

## DID YOU KNOW?

Conservation means "wise use" of our natural resources. One of the natural resources that people use is the land. This activity will challenge children to investigate the world around them, as well as encourage them to explore the many different ways people use the land.

## MATERIALS

- COLLECTION BAG FOR EACH CHILD (TO USE ON THE WALK)
- GLUE
- CRAYONS
- SCISSORS
- TAPE
- DIGITAL CAMERA OR CAMERA AND FILM (OPTIONAL)
- CHART PAPER AND MARKERS
- HEAVY PAPER, SUCH AS TAGBOARD
- CARDBOARD OR STYROFOAM MEAT TRAY FOR EACH CHILD

## ACTIVITY

1. Gather children and prepare them for a walk around the neighborhood. Discuss some of the things they might see on their walk. Record their ideas on chart paper that can be displayed after the activity. Discuss items children might collect on the walk. Be sure to talk about dangerous or inappropriate items for the children to take back to the classroom.

2. Distribute collection bags to the children and take a walk. Encourage children to collect items such as flower petals, rocks, trash, nuts, leaves, and so on. Photograph the items that cannot be taken back to the classroom. Be sure to direct the children's attention to items they might miss such as telephone poles, electric wires, garbage cans, streets, sewers, fireplugs, clouds, and so on. Remind children that flowers are for touching and smelling only, not for picking.

3. After the walk, compare items the children saw with those they predicted they would see. Ask children open-ended questions like these:
   - What things did you see and collect that were living? Nonliving?
   - What are some different ways that you saw people using the land?
   - How do you think the neighborhood looked before people lived here?
   - What animals do you think live in this environment? What animals don't? Why?

4. Encourage children to use the gathered items to create collages of their experience.

5. Create a classroom display by using children's documentations.

## ADDITIONAL ACTIVITIES

- **BLOCK**—Put out various props so children can build houses and neighborhoods in the block area during self-selected activity time. Use green paper for grass, straws and twigs stuck in clay for trees and telephone poles, spools for fire plugs, and so on. Display a map and pictures of various types of houses and buildings. Provide paper and pencils for children to create their own maps.
- **FIELD TRIP**—Visit other types of land-use areas and compare what the children see. For example, visit a farm, park, wildlife area, and downtown shopping area. Discuss the different ways people use the land in each one.
- **FIELD TRIP**—Take this same walk during the different seasons of the year. Discuss changes in the scenery, collection items, and weather. Refer back to previous documentation for a point of reference.
- **PRETEND PLAY**—During self-selected activity time put out dress-up props in the pretend play area for the types of stores or community helpers the children saw on their walk (examples include grocery store, flower shop, farmer, logger, police officer, bricklayer, construction worker, and so on).

# Tree Skin  Outside

## DID YOU KNOW?

One way trees can be identified is by their bark. Persimmon trees have rough, bumpy bark, while the sweetgum and sycamore have smooth bark. A variety of field guides are available to help you identify trees in your neighborhood. (Your state conservation agency or department of natural resources might be helpful in determining the most useful guide for your area.) This activity will encourage children to explore, compare, and contrast bark coloring and texture as well as some of the effects people have on the bark.

## MATERIALS

- **OUTDOOR AREA WITH SEVERAL DIFFERENT TYPES OF TREES**
- **BLANK NEWSPRINT OR DRAWING PAPER**
- **PEELED CRAYONS (LARGE SIZE WORKS BEST)**
- **MAGNIFYING GLASS FOR EACH CHILD**

## ACTIVITY

1. Place crayons and paper near several trees in the play area for use during outside self-selected activity time.

2. As children approach the trees, point out the textures of the bark. Show them how to replicate the texture of the tree's bark by placing drawing paper flat against the tree and rubbing it with the side of a crayon. Encourage children to make rubbings of several different trees. Discuss similarities and differences in the various rubbings.

3. Have children look for tree scars. Ask: "What causes trees to get scars?" Compare the children's scrapes and cuts to scars on a tree. Discuss the similarities between blood and tree sap.

4. Compare and contrast tree bark with people's skin. Ask open-ended questions like these: "How are tree bark and people skin alike? How are they different?" Look at colors and textures of both kinds of "skin."

## ADDITIONAL ACTIVITIES

- **BULLETIN BOARD**—Create a display by matching the children's tree rubbings with actual tree leaves or pictures of the tree.
- **OUTSIDE**—Go on a leaf rubbing scavenger hunt. Provide children with one rubbing each and challenge them to find the tree it came from.
- **SCIENCE**—Obtain a cut round (tree cookie) or a branch from someone's wood pile. Explain how the tree grows a new ring of wood each year. Encourage children to age the tree by counting the rings. Compare center rings with outside rings.

# Molly and the Forest Fire  Story

## DID YOU KNOW?

In the Midwest, forest fires occur in the spring due to leaf litter and dead annuals on the ground. These are generally ground fires and don't damage trees as heavily as crown fires in the West. Late summer and fall is the high-risk time for forest fires in the western United States due to weather conditions, decreased rainfall along with shortsighted land use, decades of fire suppression, and improper forest management (see p. 2 for further discussion). This story will help children become aware of the fact that most forest fires are caused by careless people. It will also point out some of the consequences of forest fires.

## MATERIALS

- **FALL PATTERN 1 (SEE APPENDIX D, P. 212)**
- **FELT CHARACTERS**
- **FLANNELBOARD**

## ACTIVITY

1. Read the following story to the children, placing the felt characters on the flannelboard at the appropriate points in the story.

2. Ask children open-ended questions like these:
   - Where do you think Molly will go to live now?
   - What happened to the other animals?
   - How is fire helpful to people and animals?

## ADDITIONAL ACTIVITIES

- **ART**—During self-selected activity time, put out black chalk and white paper for children to create pictures of a forest following a fire.
- **BLOCK**—Create trees from twigs and clay. Place these, along with plastic animals, in the block area during self-selected activity time. As children are playing, suggest that the forest has just caught fire and the children need to decide what to do.
- **BULLETIN BOARD**—Display pictures of forests before, during, and after fires.

Molly was a small bird,
her home was in a tree.
A forest grew around her
as far as she could see.

Big trees, little trees,
trees with holes inside
made good homes for animals,
where they could go and hide.

Molly woke up with the sun.
Each day she'd scratch and pick
and if she saw a bug or seed
she'd chomp it up real quick.

She flew up to her hole each night,
high up in the tree.
She was safe from predators
or things she couldn't see.

Molly watched the other animals
from high up in the tree.
Raccoons and deer and fox squirrels
and turkeys she could see.

The forest was their home,
like a house for you and me.

These critters would not be there,
if they didn't have those trees.

One day a man came hiking
to find a good campsite.
He started up a campfire
to keep him warm that night.

Early in the morning
the camper walked away.
He left his campfire burning
in a very careless way.

Soon the fire got bigger,
it flamed and cracked and popped.
The fire made Molly frightened.
She chirped and flapped and
  hopped.

*Woosh*, the wind came blowing.
It blew the flames about.
*Roar*, the fire grew bigger,
no one to put it out.

Smoke and flames, burning heat,
burning trees and ground.
Molly had to fly away.
She couldn't stay around.

She spent the night high in a tree,
next to a little stream.
The smoke and flames had scared her
like a very bad, bad dream.

The next day she went flying.
She flew back to her home.
The forest floor was blackened ashes
and all the brush was gone.

Bugs, seeds, acorns, leaves
were ashes on the ground.
Not a single deer or squirrel
or critter could be found.

Molly couldn't stay around
when all her food was gone.
Molly had to fly away
and find another home.

The forest soil will wash away
whenever there's a rain.
And years will pass before the woods
will be the same again.

—by John Griffin

- **FIELD TRIP**—Visit a wildlife or recreational area where campfires are permitted. Build a campfire and roast hot dogs and marshmallows. Let the children help put the fire out properly under close adult supervision.
- **GROUP**—Invite a firefighter, forester, or Smokey the Bear to visit your class.
- **OUTSIDE**—During self-selected activity time, provide firewood and buckets in the play yard for children to pretend to build campfires and put them out.
- **PRETEND PLAY**—Put out firefighter dress-up clothes for children to wear during self-selected activity time.
- **SCIENCE**—Display pieces of charred wood and magnifying glasses during self-selected activity time. Be sure to place science journals nearby for children to document their observations.
- **STORY**—Read *The Four Elements: Fire* by Maria Ruis and J. M. Parramon (see appendix A on p. 184). This simple story illustrates how fire both helps and harms people and animals.

# Hang On! 🍂 Science

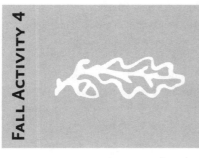

## DID YOU KNOW?

Trees help hold soil in place and keep it from washing away. This activity will help children explore another benefit of trees.

## MATERIALS

- **TUB OF SOIL**
- **STICKS AND TWIGS**
- **WATER**
- **CHART PAPER AND MARKER**

## ACTIVITY

1. Place soil in the sensory table and encourage children to experiment and play with it during self-selected activity time. Sticks and twigs can be added to represent trees.

2. As children are playing, discuss what would happen if water was added to the soil. Write their ideas on chart paper.

3. Have each child put a hand in the tub of soil and put their fingers around a handful. Ask them to pretend their arms are trees and their hands are tree roots. Together the children make up a forest of trees. A big storm is coming and the trees have to hang on. Add water to the soil, pretending it is a heavy rainstorm.

4. Ask children open-ended questions like these:
   - How do tree roots help the tree?
   - What happens to the soil when there aren't any trees?
   - What other ways are there to keep soil in one place?
   - What other plants help hold the soil?

   Document the children's thoughts and ideas and compare them with their predictions.

5. Repeat the experiment as different children visit the area and express interest.

## ADDITIONAL ACTIVITIES

- **BULLETIN BOARD**—Take pictures of the children as they are conducting the experiment. Display the pictures along with their predictions and ideas about the experiment.
- **FIELD TRIP**—Visit a site where a bulldozer has pushed over trees or a creek bank and examine the exposed roots.
- **OUTSIDE**—During outside self-selected activity time, encourage children to look for surface tree roots on the playground.

# Do Trees Get Drinks? ✤ Science

## DID YOU KNOW?

Trees get water from the ground through their roots. The water travels up "tubes" in the tree all the way to the leaves at the top. If there isn't enough water, the tree will die. This activity will enable children to see how moisture moves up the tree to the leaves.

## MATERIALS

- **KNIFE**
- **CELERY STALK WITH LEAVES (WORKS BEST WITH CELERY THAT HAS BEEN OUT OF THE REFRIGERATOR FOR SEVERAL HOURS)**
- **TWO CLEAR GLASSES OR JARS**
- **FOOD COLORING IN TWO COLORS**
- **SCIENCE JOURNALS**
- **CHART PAPER AND MARKER**
- **CAMERA (OPTIONAL)**

## ACTIVITY

1. With a small group of children during self-selected activity time, trim away the bottom part of the celery stalk and slice halfway through the center of the celery stalk lengthwise. (It takes several hours for the dye to travel through the celery, so do this activity at the beginning of the day or just before the children go home.)

2. Ask children to fill the two containers about three-quarters full of water and add enough food coloring to make a dark solution of one color for each container (red and blue work best).

3. Put the two containers next to each other and place the celery stalk so it has a cut end in the water in each jar. Ask children open-ended questions like these:
   - How do you think celery and trees are alike? How are they different?
   - What will happen to the celery? Why?

4. Ask children to predict what will happen and have them document these predictions in their science journals. Write their ideas on chart paper and photograph the various stages of the experiment.

5. Compare their predictions with their observations of the experiment.

6. Display the experiment, their predictions and conclusions, and pictures of the experiment so children can further discuss their ideas.

## ADDITIONAL ACTIVITIES

- **SCIENCE**—Add a "control" to the above experiment by placing a trimmed stalk of celery (with leaves) in a dry jar. Notice the difference between the two experiments.

- **SCIENCE**—During self-selected activity time, provide a tub with water and have children soak the leaves in it. Discuss what happens to the veins, and compare them to veins on the children's hands. Include leaves of different shapes and different vein patterns.

# Tree Puzzle  Manipulative

### DID YOU KNOW?

Trees have several different parts—the roots, trunk, branches, and leaves. Each part is important to the survival of the tree. This activity will heighten children's awareness of the various parts of the tree.

## MATERIALS

- FALL PATTERN 2 (SEE APPENDIX D, P. 213)
- FLANNELBOARD
- FELT PARTS FOR A TREE PUZZLE— ROOTS, TRUNK, LEAVES (ALL SEASONS)
- FELT FRUIT AND/OR NUTS
- FELT BIRDS, BIRD NESTS, SQUIRRELS, SQUIRREL NESTS, RACCOONS, ETC. (THESE SHOULD BE SELECTED FROM ANIMALS THAT LIVE IN TREES IN YOUR REGION OF THE COUNTRY)

## ACTIVITY

1. Place the tree puzzle near the flannelboard during self-selected activity time.

2. Allow children to put the puzzle together on their own.

3. Vary the parts of the puzzle relevant to your area of the country, as the seasons change.

4. As children play with the puzzle, discuss what they think the parts of the tree are called and what their functions might be. Ask children open-ended questions like these:
   - Why do you think trees need all of these parts?
   - How does the tree change?
   - How do animals use trees?
   - How do people use trees?

   Accept all responses and encourage children to discuss the possibilities with one another.

5. Use the fruit, nuts, birds, squirrels, nests, and any other relevant animals to encourage play and discussion about the many uses of trees.

## Ideas from Sherri's Classroom

The flannelboard is a favorite place for children to work in my classroom. However, in many programs, it is often an under-utilized area. When I make this flannelboard puzzle available, I begin with just the basic parts of the tree. As children manipulate and experiment with the tree itself, I add animals, birds, nests, trees, and so on—just a few pieces at a time. This keeps the activity interesting while providing an opportunity to discuss which tree parts are important to various animals as well as codependence of all the plants and animals.

## Additional Activities

- **Art**—Provide brown construction paper, paint and leaf-shaped sponges, tissue paper, torn construction paper, and other materials for children to create their own trees. Place materials, as well as tree pictures, in the art area during self-selected activity time. Vary the materials depending on the season and the children's ideas.
- **Music**—Play a song such as Charlotte Diamond's "What Kind of Tree Are You?" and have children move to the music.
- **Story**—Read *The Tree: A First Discovery Book* by Pascale de Bourgoing (see appendix A on p. 188). The story follows a chestnut tree as it grows and changes through the seasons. A description of other trees and ways to recognize them are also included.

# Woodworking Comparisons 🍃 Woodworking

## DID YOU KNOW?

Different trees produce wood that is different in color, texture, hardness, and odor. This activity will encourage the exploration of various types of wood and experimentation with ways people use this renewable resource.

## MATERIALS

- **SEVERAL DIFFERENT VARIETIES OF SCRAP LUMBER (SUCH AS PINE, SPRUCE, CEDAR, OAK, WALNUT, BALSA, ETC.)**
- **HAMMERS**
- **NAILS**
- **HANDSAWS**
- **HAND DRILLS**
- **RULERS**
- **PENCILS**
- **SAFETY GOGGLES**

## ACTIVITY

1. Introduce a small group of children to the woodworking area by describing each tool and demonstrating its use. Allow children to practice with close supervision. (Parents might be asked to volunteer for this activity until children become comfortable using the tools.)

2. During self-selected activity time, provide various types of wood for children to experiment with. As children work, discuss similarities and differences between the woods. Ask children open-ended questions like these:
   - Which wood is the easiest for you to saw? To drill? To hammer?
   - Where does the sawdust come from?
   - How are these woods the same? Different?

3. As children become more proficient in their use of the equipment, encourage them to plan projects and select their wood accordingly.

## Ideas from Sherri's Classroom

Woodworking is an important part of any early childhood program. This activity naturally occurred in my classroom as a result of using a variety of wood scraps in our woodworking area. I noticed children searching the wood barrel and asked what they were doing. They told me they were searching for the "soft" wood because it was easier to saw. Their search led me to be more deliberate about the types and varieties of wood I placed in the wood barrel. I have found that woodworkers in our community are often helpful in making suggestions, providing some wood scraps, and coming into the classroom to work with the children as they use the tools.

The children often talk with each other about which woods work best for which projects. For example, they have discovered that while balsa wood is easier to nail and saw, it doesn't paint as easily nor does it hold up for projects placed outside, such as bird feeders. Oak is a much harder wood for the children to work with but they really notice the grain lines and beauty of it.

## Additional Activities

- **Field trip**—Visit a lumberyard, furniture store, cabinet shop, sawmill, or musical instrument maker.
- **Group**—Invite a fiddler or a woodworking craftsperson to the classroom.
- **Nutrition**—Serve various fruits and nuts for snack and discuss what kind of tree or plant they came from.
- **Pretend play**—During self-selected activity time, put out props in the pretend-play corner for a carpenter and a tree surgeon. Have artificial fruits and nuts available for children to use in pretend cooking.
- **Story**—Read *A Carpenter* by Douglas Florian (see appendix A on p. 188). The carpenter's craft is demonstrated through simple words and eloquent pictures.

# Tree Books ⨯⨯⨯ Writing

## DID YOU KNOW?

Young children already have an abundance of information about trees. This activity will help you gain more under- standing of children's knowl- edge of trees, as well as pro- vide children an opportunity to document that information.

## MATERIALS

- **WATERCOLORS**
- **CRAYONS**
- **WATERCOLOR PAPER OR OTHER HEAVY, ABSORBENT BLANK PAPER**
- **MARKERS**

## ACTIVITY

1. During group time, ask children to share what they know about trees. Write their ideas on chart paper, including at least one idea from each child.

2. In small groups, take the children outside to an area with several trees. If an outside area is unavailable, put pictures of trees in the area where the children will be working. If trees are not readily available in your area of the country, focus the activity on plants native to your region.

3. Discuss unique features of trees with the chil- dren. Ask children open-ended questions like these:
   - What lives in trees?
   - How do trees help people?
   - How do trees help wildlife?
   - What do trees need to survive?

   Talk about how the branches come out of the tree, the different leaf structures, and the roots visible on the ground.

4. After the discussion, encourage each child to select a favorite tree and draw it with crayons. After the trees are drawn, children can use watercolors to paint their trees.

5. When paintings have dried, encourage children to write and illustrate stories on the back of their paintings. Remember, even young children can pretend to write. After completing their stories, have children read them aloud to the class. Record children's reading either on paper or by making an audio recording, and attach a transcript of children's conversations to each of their writings.

6. Bind the paintings and transcripts together and create a class book for children and families to read and enjoy together.

## ADDITIONAL ACTIVITIES

- **BLOCK**—Create trees by using twigs and clay. Set these out in the block area for children to include in their building during self-selected activity time.
- **MANIPULATIVE**—During self-selected activity time, provide pictures of animals for children to sort according to those animals that live in trees and those that do not.
- **OUTSIDE**—Adopt a tree. Make a list of children's ideas about how they can help care for their adopted tree. Some suggestions include the following: water it, mulch around its trunk, fertilize it, don't pull on the branches, put bird and squirrel feeders in it, or plant flowers under it.

- **OUTSIDE**—During self-selected activity time, encourage children to look for bird nests and other signs of wildlife that uses trees. (Remember that bird nests should not be disturbed; nests carry many diseases and may not be legally possessed.)
- **STORY**—Include a pocket and card in the class book so that it can be checked out and shared at home. Put a library card in the front of the class book so children can check it out and share at home. Make an audio recording of children reading their stories to go along with the class book.
- **WRITING**—Place small, blank books in the writing center for children to create their own tree books during choice time.

# Leaf Lotto ~ Manipulative

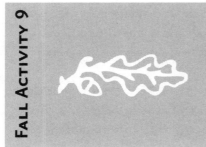

## DID YOU KNOW?

Leaves can be differentiated by their margins (edge), vein structure, and lobes. This activity will encourage children to distinguish differences in leaves.

## MATERIALS

- **TWO EACH OF AT LEAST FIVE DIFFERENT TYPES OF LEAVES**
- **CLEAR CON-TACT PAPER**
- **POSTERBOARD OR HEAVY PAPER**

## ACTIVITY

1. Make a matching game with the leaves. Attach one set to the posterboard and cover with clear Con-Tact paper. Cover the other leaves individually.

2. Place the lotto game on a table during self-selected activity time. Encourage children to match the leaf pairs. Ask children open-ended questions like these:
   - How do you decide which leaves go together?
   - How are these leaves the same? Different?

3. Add names of trees to make the game more challenging for older children.

## ADDITIONAL ACTIVITIES

- **ART**—During self-selected activity time, choose one of the following activities:
  - Provide glue and paper for children to create leaf collages with leaves collected on a field trip. Have children create a leaf sun catcher by arranging the leaves between two pieces of clear Con-Tact paper.
  - Provide real leaves and paint for children to create leaf prints.
  - Demonstrate how to make leaf foil prints by placing aluminum foil over leaves and rubbing with the side of a pencil or fingertip. Provide materials for children to create their own.
  - Demonstrate how to make leaf rubbings by placing a leaf under a piece of paper and rubbing over it with the side of a crayon. Younger children experience more success with this when leaves are taped to the table to prevent movement during the rubbing.
- **BULLETIN BOARD**—Create a bulletin board display with leaf rubbings from a variety of leaves. Under each rubbing, place a small piece of Velcro. Cover the leaves with clear Con-Tact paper and attach the opposite pieces of Velcro to the leaves. Place real leaves in a pocket on the bulletin board. During self-selected activity time, encourage children to place each leaf under its rubbing.
- **GROUP**—Pass out pairs of identical leaves to children during group time and challenge them to find someone with a matching leaf.

- **MANIPULATIVE**—During self-selected activity time, choose one of the following:
  - Create a memory or concentration game by placing leaves on cards and covering with clear Con-Tact paper. Include two of each type of leaf.
  - Cut out leaves from heavy cardboard and punch holes in them. Have children lace them with yarn for veins.
  - Provide materials for children to sort leaves by size, texture, insect damage, color, smell, and so on.
  - Cover a variety of leaves with Con-Tact paper for children to manipulate, sort, and play with.
- **OUTSIDE**—Encourage children to make wet leaf prints on the sidewalks.
- **OUTSIDE**—Pass out strips of construction paper or paint samples in a variety of fall colors or various shades of green. Challenge children to find leaves of a similar color while walking through a wooded area.
- **STORY**—Read *Red Leaf, Yellow Leaf* by Lois Ehlert (see appendix A on p. 188). This is the story of a sugar maple tree and the child who planted it.

# Nature Jar 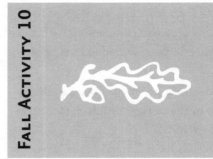 Outside

## DID YOU KNOW?

The purpose of flowers is seed making. One flower can produce one or many seeds. Seeds are the plant's way of reproducing itself. This activity will expose children to different kinds of seeds and their purpose.

## MATERIALS

- **NATURAL AREA WITH MANY TREES AND WEEDS**
- **LARGE, CLEAR JAR**
- **MAGNIFYING GLASSES**
- **SCIENCE JOURNALS**

## ACTIVITY

1. During outside self-selected activity time, ask children to gather seeds, berries, and other fruits to put into the jar. Emphasize that none of these are for eating and should not be put into their mouths. Be certain that children do not collect poison-ivy berries.

2. Have the collection jar available for children to explore, sort, compare, and match the items. Provide science journals for children to document their observations.

3. As children work, talk about the purpose of seeds and how seeds come in pods, nuts, or fruits. Ask the children open-ended questions like these:
   - Why do plants make seeds?
   - How are seeds dispersed or scattered?
   - How do people use seeds?
   - How do animals use seeds?
   - What foods do people eat that are seeds?

## IDEAS FROM SHERRI'S CLASSROOM

One of the discoveries we made during this experience occurred while examining the seeds we collected. The children found that in addition to the seeds they put in the jar, they also collected seeds with their clothing and were curious as to why some seeds stuck to them and others did not. The seeds that stuck to their clothing were examined carefully under magnifying glasses. The hooks on the seeds looked just like the hooks on Velcro. After a little research, we found that the hooks on seeds like these were actually the inspiration behind Velcro. The children were amazed that something in nature inspired something created by people and so important in their lives.

## ADDITIONAL ACTIVITIES

- **ART**—During self-selected activity time, provide materials for children to make seed collages.
- **NUTRITION**—Plan a "seed of the day" snack, serving a different type of seed each day (such as sunflower seeds, pumpkin seeds, peanuts, popcorn, and so on).
- **OUTSIDE**—During outside self-selected activity time, encourage children to collect seeds and berries from weeds and trees. Compare with commercial birdseed. Put the collected birdseed in a birdfeeder and see what the birds prefer. Write the children's observations on a chart.
- **SCIENCE**—During self-selected activity time in the classroom, display the nature jar and let the children empty it and match the seeds and fruits. Add a scale to the area so children can explore and compare weights of the various items. Be sure to have their science journals available.

# I'm a Little Milkweed Cradle 🐛 Music

## DID YOU KNOW?

There are many different ways for seeds to be dispersed or scattered. Frequently, the wind helps. This song and activity will familiarize children with one type of seed dispersal and the role of wind.

## MATERIALS

- **ONE MILKWEED POD PER CHILD**
- **OTHER TYPES OF SEEDS**
- **MAGNIFYING GLASS FOR EACH CHILD**

## ACTIVITY

1. Gather children outside in a group. Give each child a milkweed pod and encourage children to examine the pods while using the magnifying glasses.

2. Encourage children to open the pods and disperse the seeds in any way they choose. Have them experiment with other types of seeds.

3. Ask children open-ended questions like these:
   - How do you think seeds get from one place to another?
   - How might these seeds travel if there wasn't any wind?
   - How long can you keep your milkweed seeds in the air?

4. Teach children the following song to the tune of "I'm a Little Teapot":

   I'm a little milkweed cradle you see. (*Cup hands together.*)
   I have baby seeds in me. (*Peek inside.*)
   Open me up and hold me high. (*Open hands and hold high.*)
   Blow, blow wind and my seeds fly. (*Blow on hands and wiggle fingers.*)

## ADDITIONAL ACTIVITIES

- **ART**—During self-selected activity time, provide materials for children to make collages with glue, milkweed pods, seeds, and paper (or a Styrofoam meat tray).

- **LARGE MOTOR**—Provide children with several types of seeds that travel with the wind (such as milkweed, dandelions, cattails, and so on). Challenge children to blow on the seeds and see how long they can keep the seeds in the air.

- **OUTSIDE**—Challenge children to look on the playground or in a park for other seeds that travel with the wind.

- **PRETEND PLAY**—Encourage children to pretend to be seeds and imagine how they might be dispersed.

- **SCIENCE**—Place milkweed pods in the science area for children to examine with magnifying glasses, weigh, or count the seeds during self-selected activity time. Have their science journals available. Provide soil, pots, and water for children to plant milkweed seeds if they so choose. Document the process and growth with photographs or drawings.

# The Harvest ✎ Story

## DID YOU KNOW?

Every year farmers across the nation test their soil, as well as plant, fertilize, cultivate, and harvest their crops. Although most crops are harvested in fall, some crops, such as wheat, are harvested in the spring or summer. This story will expose children to how farmers grow their crops, the types of equipment involved, and what happens to the crops after harvest.

## MATERIALS

- **FALL PATTERN 3 (SEE APPENDIX D, PP. 214–216)**
- **FELT CHARACTERS**
- **FLANNELBOARD**

## ACTIVITY

1. Read the following poem to the children, placing the felt pieces on the flannelboard at appropriate places in the story.

2. Ask children open-ended questions like these:
   - What other crops do farmers grow?
   - How does the cornfield help wild animals?
   - If Farmer Fred didn't have animals, what could he do with his corn?

## ADDITIONAL ACTIVITIES

- **BLOCKS**—During self-selected activity time, provide toy farm equipment in the play area, using sewing thread spools or vegetables for children to harvest.
- **BULLETIN BOARD**—Display pictures of various farm equipment and encourage discussion about their use.
- **FIELD TRIP**—Visit a farm, farm implement company, grocery store, farmer's market, or feed store.
- **LARGE MOTOR**—Make beanbags in the shape of fruits and vegetables. During self-selected activity time, provide a basket for children to harvest the crop.

Farmer Fred went to his field.
He got a little dirt.
The laboratory tested it
before he started work.

The laboratory wrote him.
They said, "Your soil is fine.
But if you want a good corn crop,
add a little lime."

Fred drove his truck and got some lime
as quick as he could go.
He spread it on his cornfield
'til the soil was white as snow.

Fred got his plow and tractor
and plowed it like he should.
Then pulled his disc around the field
and mixed the soil real good.

He put seeds in his planter.
You need those to make corn grow.
Straight up and down his tractor went
and planted them in rows.

Farmer Fred went to his house.
And soon the rain and sun
made little corn plants in the field.
The growing had begun.

Each day the plants got bigger.
They all grew straight and green.
Very soon small ears
of corn could be seen.

The ears were wrapped in green leaves.
Inside them corn will grow.
There were ears on every plant,
in every single row.

Soon the plants were very tall.
The growing up was done.
All the leaves on every plant
turned brown in the sun.

The corn plants all died,
but they didn't fall down.
Each plant stood straight and held
   the ears
high off the ground.

Farmer Fred looked at his field.
"It's time to get to work!"
He jumped into his combine.
It started with a jerk.

Down the rows the combine went.
Click, click, the corn came down.
The plants went through the combine
and out onto the ground.

Clack, clack went the combine.
It scraped seeds off every ear.
Up a pipe the seeds went
to a big box in the rear.

Farmer Fred took the corn,
he put it in a bin.
When he fed it to the pigs
they'd look at him and grin.

He fed it to his milkcows.
He fed it to his sheep.
And through the long cold winter
they all had corn to eat.

—John Griffin

- **MANIPULATIVE**—Using pictures of fruits and vegetables, create a memory or concentration game for children to play during self-selected activity time. These same cards can be used for a fruit and vegetable sorting game.
- **NUTRITION**—Make vegetable soup for snack or lunch and have children do the preparation. Discuss where the vegetables came from.
- **OUTSIDE**—Observe farmers harvesting their crops.
- **PRETEND PLAY**—During self-selected activity time, set up a produce stand using artificial fruits and vegetables, baskets, signs, paper for making lists, money, and a cash register.
- **PRETEND PLAY**—Provide overalls, caps, bandannas, and buckets in the pretend play corner for children to pretend to be farmers during self-selected activity time.
- **STORY**—Read *Eating the Alphabet* by Lois Ehlert (see appendix A on p. 187). This book looks at fruits and vegetables from A to Z.
- **WRITING**—During self-selected activity time, provide magazine pictures, glue, markers, paper, and staplers for children to create food books of all the kinds of food grown in the United States.

# Something Corny ✺ Science/Math

## DID YOU KNOW?

Corn is a plant that stores its food in its seed. Many animals and people like to eat corn. There are several different ways to preserve corn, such as canning, freezing, and drying. This activity will enable children to explore the results of different methods of preserving foods.

## MATERIALS

- **POPCORN**
- **FROZEN CORN**
- **CANNED CORN**
- **CREAM-STYLE CORN**
- **HOMINY**
- **FIELD CORN**
- **FRESH SWEET CORN**
- **CORNMEAL**
- **TOOTHPICKS**
- **CHART PAPER AND MARKER**

## ACTIVITY

1. Set up a tasting center using toothpicks and various types of corn. Select only a few types of corn for younger children and provide more variety for older children. Ask: "How are the various types of corn the same? How are they different?"

2. Encourage a discussion about the various types of corn preservation. Ask children open-ended questions like these:
   - Which method of preservation do you think is best for people? Which method is the best for animals?

   On chart paper, list children's responses and have them vote on the type of corn they like best.

3. Encourage children to visit the tasting center during self-selected activity time. Ask them to sign their names under the type of corn they like best.

4. During group time, look at the voting chart and see which type of corn was preferred by the most children in the class.

## IDEAS FROM SHERRI'S CLASSROOM

This activity came from a discussion the children had about different kinds of corn. Some of them talked about corn on the cob being their favorite, while others preferred it out of the can. As a result of their discussion, a class taste test was set up. Not only did it promote discussion about favorite ways corn is prepared, but also various methods for saving the corn to eat and enjoy later.

## ADDITIONAL ACTIVITIES

- **MANIPULATIVE**—Put out clean rocks and dry corn for children to experiment with grinding their own cornmeal during self-selected activity time. Use their cornmeal to make cornbread for snack.
- **NUTRITION**—Make popcorn for snack.
- **NUTRITION**—Make Johnny cakes by mixing cornmeal and water together and making patties to fry. Try them for snack.
- **PRETEND PLAY**—During self-selected activity time, place empty food containers from dried, frozen, and canned foods in the housekeeping area.
- **STORY**—Read *Corn Is Maize: The Gift of the Indians* by Aliki Brandenberg (see appendix A on p. 182). This simple story tells the history of corn from its discovery by Native Americans to present day importance.

# Apple Pizzas Nutrition

FALL ACTIVITY 14

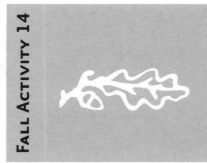

### DID YOU KNOW?

Apples are harvested throughout the United States. Many children have the opportunity to experience their growth and harvest firsthand. This activity will reinforce the children's knowledge and provide an opportunity to enjoy one of our nation's harvests.

## MATERIALS

- PEELED AND SLICED APPLES
- ONE-HALF ENGLISH MUFFIN OR BAGEL PER CHILD
- MARGARINE
- SUGAR
- CINNAMON
- SHREDDED CHEESE (OR LET CHILDREN GRATE CHEESE)
- ALUMINIUM FOIL
- PERMANENT MARKER

## ACTIVITY

1. Tear enough small pieces of aluminium foil for each child to have one. Place these and the English muffins or bagels at the end of a table.
2. Set out the margarine, apples, sugar, and cinnamon. Place the shredded cheese last.
3. During self-selected activity time, encourage children to individually assemble their pizzas. First, place an English muffin or bagel on a piece of foil. Spread margarine over the muffin or bagel. Add a few slices of apple. Sprinkle with sugar and cinnamon. Finally, sprinkle lightly with shredded cheese.
4. Encourage children to write their name on the foil with the permanent marker.
5. Place the pizzas (foil and all) on a cookie sheet and bake at 350°F until the cheese melts. Serve for snack.
6. Ask children open-ended questions like these:
   - How do you fix apples at your house?
   - Where do apples come from?

## Additional Activities

- **Field trip**—Visit an apple orchard.
- **Large motor**—Teach the children to play "Worm through the Apple." The "Apple" stands with legs wide. The "Worm" tries to crawl through the "Apple" without knocking it off the tree. Discuss other insects and animals who like to eat apples.
- **Manipulative**—During self-selected activity time, allow children to help prepare apples and count the seeds in each one. Provide scales and tape measures for children to weigh and measure the apples. Make a chart to record the results.
- **Manipulative**—Under close adult supervision, allow children to experiment using an apple peeler, slicer, and corer. Use the peeled and sliced apples for your apple pizzas.
- **Nutrition**—Provide opportunities for children to make applesauce, apple butter, apple pie, apple jelly, or apple juice.
- **Science**—Set up an apple taste center using several different kinds of apples. During self-selected activity time, encourage children to taste and compare flavors, textures, colors, sizes, and aromas. Write their comments on a chart to determine the favorite.
- **Science**—During self-selected activity time, encourage a small group of children to examine the oxidation of an apple by putting one apple slice in lemon juice and leaving the other in the open air. Be sure to document their predictions before beginning. Compare, taste, and record the results. Photographs of the process make an interesting display and allow children to revisit the experience.
- **Story**—Read about Johnny Appleseed.

# Animal Harvest 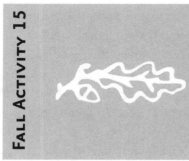 Nutrition

## DID YOU KNOW?

Farmers harvest their farm animals as well as their plants. Many of the foods we eat come from farm animals. This activity will help children identify some of the animal products people get from farmers.

## MATERIALS

- BACON
- EGGS (1 PER CHILD)
- MILK
- ELECTRIC SKILLET
- MIXING BOWL
- WIRE WHISK OR ROTARY BEATER
- WOODEN SPOON
- PANCAKE TURNER

## ACTIVITY

1. Set up a table in the classroom where the children, without getting too close, can watch the bacon frying in the electric skillet.

2. As the bacon fries, discuss bacon's origin. Talk about other products hogs provide, such as sausage, ham, pork chops, and so on. Talk about how the bacon changes and the aroma it produces as it fries.

3. Encourage each child to crack an egg and empty it into the mixing bowl. Take turns beating the eggs for scrambled eggs. Add milk and scramble the eggs in the skillet.

4. As the eggs are cooking, discuss where eggs come from and other products farmers get from chickens.

5. Serve milk with the bacon and eggs for snack or lunch. As the children eat, talk about other products that come from animals. Discuss various farm animals and wild game, such as deer, fish, ducks, turkeys, rabbits, and so on, and how they help the farmer earn money to buy clothing, food, and other essentials. Ask children open-ended questions like these:
   - How does your family make money to buy clothing, food, and other essentials?
   - What farm products do you use?

## IDEAS FROM SHERRI'S CLASSROOM

This activity is an important discussion starter for helping children understand where food comes from. Although adults sometimes dislike talking about animal food products, children deserve the respect we show when sharing this information. However, it is important to listen to the children and not provide more information than they are ready for. The concept that hamburger comes from cows, bacon comes from hogs, and eggs come from chickens, is more than likely as much as young children will be capable of understanding.

The concept that most working adults earn money for their families is also important for this experience. Relating this idea to things their parents do to earn a living would be an important extension of the idea. Talking about farmers raising animals and crops to feed and clothe their families is information young children can understand.

A beginning understanding of both of these concepts helps children place value on food and the products we harvest from plants and animals. In today's culture of consumerism, the idea that people work and use natural resources to make money for the things they buy is vital to future wise use of natural resources.

## ADDITIONAL ACTIVITIES

- **GROUP**—During meals and snack, discuss where the food was grown and harvested.
- **MANIPULATIVE**—Make plant and animal product puzzles for children to match pictures of plants and animals from your area of the country with pictures of the products people use from them.
- **NUTRITION**—Make butter by letting children shake whipping cream in small jars. It takes about five minutes of continuous shaking until it separates. Encourage children to taste the buttermilk. Wash the butter and add salt. Serve with crackers or biscuits.
- **PRETEND PLAY**—Display pictures of various farm products in the pretend-play corner. During self-selected activity time, put out various food props, such as plastic eggs (with insides made of felt), empty milk cartons, mounted cardboard pictures of fried eggs, hamburgers, bacon, and so on. Add pails for milking cows and baskets for gathering eggs and farm clothes for dress-up.
- **SCIENCE**—During self-selected activity time, display wool, feathers, honeycomb, wood, and leather articles for children to explore. Be sure to provide magnifying glasses, scales, and science journals for children to examine materials and document their discoveries.
- **STORY**—Read *Pancakes for Breakfast* by Tomie dePaola (see appendix A on p. 182). This picture book illustrates a little old lady who wants pancakes for breakfast. Before she can make her pancakes she must gather eggs, milk the cow, churn the butter, and buy maple syrup.

# Cave Life 🙰 Art

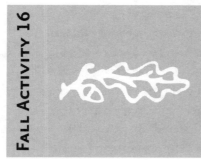

## DID YOU KNOW?

Caves provide a special kind of habitat for many forms of wildlife. This activity will help children become more aware of the special features of caves and their inhabitants.

## MATERIALS

- **LARGE ROLL OF BROWN PAPER**
- **MARKERS**
- **TAPE**
- **PAINT**
- **BRUSHES**
- **SCISSORS**
- **PICTURES OF CAVES**

## ACTIVITY

1. During group time, ask children if any have ever visited a cave. Discuss their responses, what they saw or felt in the cave. Explain that they will be creating their own cave. Brainstorm materials they might need for the project.

2. During self-selected activity time, display the cave pictures near the cave site and provide requested materials for children to work on their cave. Drape brown paper around the space for the interior of the cave.

3. Ask children open-ended questions like these:
   - How is the cave habitat different from other habitats we have talked about?
   - What plant life do you think would live in a cave? Why?
   - What animals might live in a cave? Where in the cave would they live?
   - Where can we go to find out more information about caves?
   - How do you think caves are formed?
   - How could we get bats to live nearby?

4. Encourage children to research cave formations and animals that live in caves, and then make and place these throughout the cave. This project could continue over several days or weeks.

This activity came about at the end of a summer study of caves. When we first began the study, I didn't know much about caves. Although I was a little nervous about embarking on a study of a topic I knew so little about, I learned a great deal along with the children. That summer, many of the families involved in the program included cave visits in their summer vacation plans. Building the cave was a way for the children to put their knowledge of caves together—they offered tours to the community. Their role as tour guides allowed them to demonstrate their acquired knowledge about caves. It was also a wonderful way to assess their understanding.

## ADDITIONAL ACTIVITIES

- **ART**—During self-selected activity time, provide an assortment of materials for children to create small-scale models of caves.
- **FIELD TRIP**—Visit a cave. Be sure to research the cave prior to your visit and have children wear bike helmets with flashlights affixed to the top with duct tape.
- **GROUP**—Invite a local caving group to share information with the children.
- **MANIPULATIVE**—During self-selected activity time, encourage children to sort animal pictures by those that live in caves and those that don't.
- **STORY**—Read *One Small Square: Cave* by Donald M. Silver (see appendix A on p. 180). This informational picture book explains the various parts of the cave and discusses creatures that live there.

# Worth Their Weight ✎ Science

## DID YOU KNOW?

Bats make an important contribution to the balance of nature but are often misunderstood by humans. Bats find their way in the dark through a special sense called *echolocation*. The bat makes noises as it flies. These sounds bounce off objects as echoes. These echoes enable the bat to know its location and determine objects in the environment around it. Most North American bats eat mosquitoes, moths, and other insects. In fact, most bats eat their weight in insects each day. This activity will help children understand one of the ways bats help people.

## MATERIALS

- **BAT REPLICA (PLASTIC, SPONGE, OR PLUSH)**
- **SMALL PLASTIC INSECTS**
- **BALANCE SCALE**
- **PICTURES OF BATS INDIGENOUS TO YOUR AREA**

## ACTIVITY

1. During self-selected activity time, place the scale, insects, and bat replica in the science area.

2. Explain that bats eat their weight in insects each day. Challenge children to place the bat replica on one side of the scale and see how many insects it takes to balance the scale.

3. Ask children open-ended questions like these:
   - How do bats know where to find insects?
   - What do you think would happen if there weren't any bats to eat the insects?
   - Where do you think bats live?
   - How do you think bats find their way in the dark?
   - How are bats like other mammals? How are they different?

## ADDITIONAL ACTIVITIES

- **ART**—During self-selected activity time, provide materials for children to sponge-paint bats on black paper. Provide pencils for children to draw insects for the bats to eat.
- **GROUP**—Conduct echo experiments with the children.
- **LARGE MOTOR**—Set up a simple obstacle course for children to go through blindfolded during self-selected activity time. As children are moving through the course, discuss how bats move in the dark.
- **MUSIC**—Play a song such as Jan Syrigos's "Hairy Not Scary" and have children move like bats to the music.
- **NUTRITION**—Cut bat shapes from fruit leather for snack. Discuss similarities between the texture of the fruit leather and bat skin.
- **STORY**—Read *Bat Loves the Night* by Nicola Davies (see appendix A on p. 184). This simple story about a pipistrelle bat includes numerous bat habits and facts.

# I'm a Little Fox Squirrel 🐾 Music

## DID YOU KNOW?

Squirrels store most of the nuts they collect in the ground. This not only gives squirrels something to eat in winter, but also aids in seed dispersal. This song will help children become aware that squirrels eat nuts and store them in the ground.

## MATERIALS

* **PICTURE OF A FOX SQUIRREL**

## ACTIVITY

1. Show children the picture of the fox squirrel and teach them the following song to the tune of "I'm a Little Teapot."

   I'm a little fox squirrel, red and brown. (*Put hands behind back like a squirrel tail.*)
   I eat nuts that I have found. (*Pretend to eat nuts.*)
   I pick them up and bury them deep. (*Pretend to pick up nuts and bury them.*)
   When winter time comes I'll have lots to eat. (*Rub tummy.*)

2. Ask children open-ended questions like these:
   * What happens when the squirrel forgets where he buried the nuts?
   * What else might a squirrel eat?
   * Where do you think squirrels live? Is it different in the city and in the country? How?
   * What kind of sound do you think a squirrel makes?

## ADDITIONAL ACTIVITIES

* **BLOCK**—Add play food (such as fruits or nuts) for plastic animals to eat or store during self-selected activity time.
* **FIELD TRIP**—When winter snows come or the ground is muddy enough to show footprints, visit a wooded area or park and look for signs of where squirrels dug up nuts buried in the fall.
* **GROUP**—Discuss where other animals store their food.
* **STORY**—Read *Nuts to You!* by Lois Ehlert (see appendix A on p. 184). This wonderfully illustrated story is about a gray squirrel in the city. Many squirrel facts are included at the end of the story.

# See How the Turkey Grows ⚬⚬⚬ Story

## DID YOU KNOW?

Wild turkeys are found throughout the United States. The wild turkey has many enemies, and it must escape predators its entire life. This activity will expose children to the life cycle of the wild turkey.

## MATERIALS

- FALL PATTERN 4 (SEE APPENDIX D, PP. 217–218)
- FELT CHARACTERS
- FLANNELBOARD

## ACTIVITY

1. Read the following poem to the children, placing the characters on the flannelboard at the appropriate point in the story.

2. Ask children open-ended questions like these:
   - What would the fox have eaten if it hadn't been able to catch a turkey?
   - Why do you think that turkeys have so many babies at one time while people usually have one baby at a time?
   - What did the hunter do with the turkey?
   - What might have happened to Jane if the hunter hadn't harvested one of the turkeys?

## ADDITIONAL ACTIVITIES

- FIELD TRIP—Discuss turkey season and that many states have a large turkey population that makes turkey hunting possible under controlled conditions. Visit a wildlife check station.

## IDEAS FROM SHERRI'S CLASSROOM

Many people with long experience in conservation ethics have remarked on how this short, simple story is the best expression of conservation, as it relates to wildlife management, they have ever heard. The story includes elements of habitat, life cycle of animals, habits of various species, seasonal impact on animal population, food chains, and predation, as well as responsible and ethical harvest. These all work together to insure a wise and balanced use of any resource and in the case of this story, a continuation of the species. The story simplifies these complicated concepts in a way children can easily understand and provides a powerful foundation for their future wise use of natural resources.

---

Momma turkey made her nest
in some low brush and some grass.
She had to lay some eggs soon,
so she had to work real fast.

She clicked and kelped. She laid
    six eggs.
She sat on them for weeks.
Until one day Momma turkey heard
five little "peeps."

One egg didn't hatch.
It got wet from the rain.
But five turkeys made it,
and one was named Jane.

Jane and the others
followed Momma through the weeds.
They were very watchful
as they chomped down bugs and
    seeds.

"Sssss" went a snake,
who was lying in the grass.
He chomped one of Jane's brothers,
he chomped him down real fast.

Four little turkeys left,
more watchful than before.

Along came a red fox
and chomped up one more.

Jane and the other two,
they would scratch and dig,
uncovering bugs and acorns,
they began to grow real big.

Soon it was fall,
and the leaves turned brown.
Along came a hunter,
walking all around.

He waited for the turkeys.
*Boom*, went his gun.
He saw all the turkeys,
but he only shot one.

Then the winter snow came.
It covered up the ground.
It covered up the turkeys' food,
so none could be found.

The winter got Jane's mother.
It got her brother too.
But Jane scratched up some acorns
so she made it through.

Now only Jane was left.
She was all alone.
But soon the warmth of spring came
and she was nearly grown.

One day she heard a gobble.
She gave a kelping sound.
Here came a turkey,
strutting all around.

He was a handsome turkey!
And he had a fine song!
So Jane thought she would stay
    with him,
all day long.

Jane went and made a nest.
And late in the spring
she laid six eggs and settled down
to see what time would bring.

She clicked and kelped, she sat
    and sat.
She sat on them for weeks.
Until one day Jane heard
six little "peeps."

—John Griffin

# I'm a Very Fine Turkey  Music

## DID YOU KNOW?

Wild turkeys live in forests and roost in trees at night. They eat insects and seeds. This activity will help children understand that there are turkeys that are wild as well as those that live on farms. It will also introduce the concept of "roost."

## MATERIALS

- **PICTURE OF A WILD TURKEY**
- **PICTURE OF A DOMESTIC TURKEY**
- **CHART PAPER AND MARKER**

## ACTIVITY

1. During group time, discuss with children the differences and similarities between wild turkeys and domestic turkeys raised by farmers. Ask:
   - How are wild turkeys the same as turkeys raised by farmers? How are they different?

   Write their responses on chart paper.

2. Teach children the following song:

3. Once children are familiar with the song, ask: "What does it mean to roost in the trees?"

## IDEAS FROM SHERRI'S CLASSROOM

Inviting a hunter to share turkey calls with the class is often an informative as well as fun experience for the children. Hunters can often perform several different turkey calls, including the female or hen's kelp. Hen turkeys kelp, and the tone of the kelp communicates different information to the poults, or young turkeys. The young poults aren't able to kelp yet, so they whistle at the same pitch. The hen and poults communicate to one another with this series of kelps. An experienced hunter can make a variety of these sounds and teach some to the children. Encouraging the children to be poults looking for food in the outdoors and then hurrying to the hen when the distress kelp is given provides them with better understanding of how the hen turkey cares for her young.

## ADDITIONAL ACTIVITIES

- **FIELD TRIP**—Visit a turkey farm to view domestic turkeys. Discuss differences between wild and domestic turkeys. Discuss how turkeys in the grocery store look compared to live ones.
- **GROUP**—Invite a hunter to come into the classroom, reproduce turkey calls, and, if possible, bring in a legally harvested wild turkey.
- **LARGE MOTOR**—Teach children to play "Catch a Turkey." Several children are the hunters, the rest are wild turkeys. The hunters are sleepy so they take a nap (cover eyes while the turkeys hide). The hunters wake up hungry and want some wild turkey for supper, which means it's time to go hunting. As the hunters tag the turkeys, both go and sit where the hunters were sleeping. Only one turkey is permitted for each hunter. To help the hunters find the turkeys, they can gobble softly.
- **SCIENCE**—Bring in domestic turkey feathers and wild turkey feathers. During self-selected activity time, display them for children to compare and match them with pictures of the animals. Be sure to provide magnifying glasses and science journals for children to explore and record their thoughts and ideas.
- **STORY**—Read *All about Turkeys* by Jim Arnosky (see appendix A on p. 179). This nonfiction picture book addresses many facts about turkeys in an interesting way.

# Winter

WINTER IS A TIME OF ANXIOUS ANTICIPATION for children and adults alike. People enjoy outdoor winter sports, such as skiing, sledding, skating, and birdwatching. All of these activities are examples of people using natural resources.

The winter climate changes throughout North America have an impact on seasonal changes for people, plants, and animals. During the winter months, in many regions of the country, snow blankets the outdoors. The snow provides an insulating effect that protects some plants and animals.

The Southeast United States experiences their dry season, making alligators easier to spot as they move between wetland areas in search of better habitat. On the West Coast, winter is the height of the rainy season causing mudslides as well as the emergence of slugs, mushrooms, ferns, and other debris-eating organisms.

Fur-bearing mammals respond to the colder temperatures by growing dense underfur and long guard hairs. Fur resources are harvested through trapping, which serves as an effective population management tool.

The learning events in the winter section of *My Big World of Wonder* take advantage of the many naturally occurring phenomena. Topics include the following:

- Birds and their characteristics
- Mammals in winter
- Hibernation
- Animal signs
- Weather
- Energy use

# Birdfeeders ❦ Science

## DID YOU KNOW?

There are many different species of birds. Creating birdfeeders and observing the birds that feed at them can be a fun and educational experi- ence for young children. If a program of winter bird feeding is initiated, it should be con- tinued throughout the winter months.

## MATERIALS

- **CARDBOARD TUBE FROM A TOILET PAPER ROLL**
- **PEANUT BUTTER**
- **BIRDSEED**
- **HOLE PUNCH**
- **YARN**
- **PLASTIC KNIVES**

## ACTIVITY

1.  Prepare the cardboard tubes by punching a hole in the top and stringing yarn through the hole. This provides a loop for hanging the feeder.

2.  Put out the prepared cardboard tubes, peanut butter, plastic knives, and a flat container of birdseed on a table during self-selected activity time.

3.  Encourage children to spread the peanut butter on the cardboard tubes and then roll the feeder in birdseed. Ask them open-ended questions like these:
    - Which seeds do you think the birds will like best?
    - Where do you think is the best place to hang your feeders?
    - What kinds of birds do you think will visit your feeders?

4.  Hang the feeders outside a classroom window, slightly above or next to a branch where birds can reach it and to keep away neighborhood cats. The birds should soon be regular visitors to your school.

5.  Remove the birdfeeder once all the food has been eaten.

## ADDITIONAL ACTIVITIES

- **BULLETIN BOARD**—Put up labeled pictures of birds seen at your feeders. Chart the number of sightings for each kind.
- **GROUP**—Talk with the children about things to look for at the feeder. Observe and record the following:
  - What kinds of food do the birds prefer?
  - Where do the birds eat?
  - Who stays the longest at the feeder?
  - What time of day is the feeder the most popular?
  - Which bird is the most aggressive at the feeder?
  - Which kind of feeder do the birds prefer?
  - Which feeder location do the birds seem to prefer? Why?
- **MANIPULATIVE**—Use pictures or stickers of birds to play a memory, concentration, or domino game during self-selected activity time.
- **OUTSIDE**—Put up several different kinds of feeders and see if the birds prefer any of the following:
  - Fill half a grapefruit peel or half a coconut with birdseed and peanut butter.
  - Create a holiday tree for the birds by stringing popcorn and cranberries and hanging different feeders on it.
  - Stuff a pinecone with peanut butter and birdseed.
  - Cut doors in a plastic milk jug and fill the bottom with birdseed.
  - Hang mesh or wire bags of suet or fat in the trees to attract woodpeckers.
- **SCIENCE**—Encourage children to make a birdfeeder for their home. Provide a journal for them to compare the birds they see at home with those seen at school.
- **SCIENCE**—During self-selected activity time, provide materials for children to plant birdseed.

Follow its progress over several days. Document the experience with photographs and transcriptions of the children's discussions.
- **STORY**—Read *Night Tree* by Eve Bunting (see appendix A on p. 179). This is the story of a family who makes their annual holiday pilgrimage to decorate a particular tree for wildlife.
- **WOODWORKING**—Display pictures and plans for birdfeeders and birdhouses in the woodworking area.
- **WRITING**—Have science journals available for children to document their bird observations.

# Pick a Beak   Manipulative

## DID YOU KNOW?

The many different species of birds eat a variety of things. Some are insect eaters, some are seed eaters, some are birds of prey, and others are scavengers. Each bird has a beak that is adapted to its habitat and the food it needs for survival. This activity will help children gain an understanding of how the bird's beak enables it to eat.

## MATERIALS

- PLIERS (SEED EATERS)
- TONGS (FISH EATERS)
- TWEEZERS (INSECT AND WORM EATERS)
- SLOTTED SPOON (AQUATIC PLANT EATERS)
- STRAW (NECTAR EATERS)
- SUGARWATER IN A BOTTLE (NECTAR)
- PARSLEY IN A BOWL OF WATER (AQUATIC PLANTS)
- PLASTIC INSECTS OR RAISINS IN A BOTTLE (INSECTS)
- LARGE PLASTIC FISH OR BANANA (FISH)
- ACORNS
- SUNFLOWER SEEDS
- PICTURES OF VARIOUS BIRDS (BIRDS INDIGENOUS TO YOUR REGION OF THE COUNTRY WILL BE MOST MEANINGFUL TO CHILDREN)

## ACTIVITY

1. Place the materials, along with the pictures of the birds, in the manipulative area for children to explore during self-selected activity time.

2. Encourage children to try the different utensils for picking up the food.

3. Compare the utensils to the bird beaks shown in the bird pictures. Ask children open-ended questions like these:
   - How is this utensil similar to a bird beak?
   - Which utensil works best for each food?
   - What do you think would be the habitat of a bird with this kind of beak?

4. Document children's ideas about the bird beaks and display their findings along with the bird pictures.

## ADDITIONAL ACTIVITIES

- **MANIPULATIVE**—Make a chart of various foods that birds eat, such as insects, seeds, small mammals, worms, and so on. During self-selected activity time, encourage children to sort bird pictures by what they eat.
- **NUTRITION**—Serve dry, crispy rice cereal for snack and pretend that it's birdseed. Let the chil-

## IDEAS FROM SHERRI'S CLASSROOM

This learning experience came about when children asked questions about why some birds ate at the feeders and others did not. A discussion about food and the variety of beaks resulted. The children talked about materials we could use to simulate various bird beaks. One of the important considerations when talking about habitat is how the plant or animal has adapted to take full advantage of the resources available in a particular habitat. This activity led the children to come up with their own conclusions. The bird's beak is adapted to the type of food it eats and how it obtains the food. For example, woodpeckers and flycatchers are both insect eaters and have tweezer-like beaks for grabbing insects. However, the woodpecker's beak is long and strong for drilling as well as grasping insects, while the flycatcher's beak is wider and shorter for catching insects in flight.

dren try to eat like birds. Discuss how well birds are suited for eating with their beaks but people are not so well suited with their mouths.

- **STORY**—Read *Beaks!* by Sneed B. Collard III (see appendix A on p. 179). This simple nonfiction book describes various types of bird beaks and how birds use them to hunt, gather food, and eat.

# Bird Puzzles ❧ Manipulative

## DID YOU KNOW?

Birds' feet and beaks vary according to habitats and the types of food they eat. Some birds need webbed feet for swimming, while others need claws for holding onto trees. Long, skinny beaks help birds get insects out of trees, while short, strong beaks are good for cracking seeds. This activity draws the children's attention to the color and size of birds and to differences in their feet and beaks.

## MATERIALS

- **WINTER PATTERN 1 (SEE APPENDIX D, P. 219)**
- **FLANNELBOARD**
- **BIRD PICTURES**

## ACTIVITY

1. Display the bird pictures near the flannelboard and felt bird pieces during self-selected activity time.

2. Encourage children to experiment with placing different feet and beaks on the birds. Ask them open-ended questions like these:
   - Why do you think that bird needs those kind of feet?
   - What does the woodpecker eat?
   - How does its beak help?
   - How do its feet help?
   - In what kind of habitat do you think that bird lives?

This activity takes "Pick a Beak" (see winter activity 2) one step futher by encouraging children to consider how the bird's beak and feet work together to aid it in gathering food, finding or building shelter, escaping predators, and caring for their young.

## ADDITIONAL ACTIVITIES

- **BULLETIN BOARD**—Create a display of bird beaks, feet, the foods they eat, and their habitats.
- **MANIPULATIVE**—During self-selected activity time, have bird pictures and pictures of various habitats available to the children for matching birds with their habitats. Invite children to sort bird pictures by type of beaks or feet.
- **MANIPULATIVE**—Make more bird puzzles for children to experiment with by laminating and cutting apart pictures of birds.
- **PRETEND PLAY**—Encourage children to pretend they are birds. Challenge them to stand and feed like different kinds of birds with different types of feet and beaks.
- **SCIENCE**—Challenge children to observe the different ways birds perch and eat at the feeders in the play yard. Encourage children to document their observations in their science journals.
- **STORY**—Read *About Birds: A Guide for Children* by Cathryn Sill (see appendix A on p. 180). This book provides an introduction to birds and their unique characteristics.

# Stuffed Birds  Art

## DID YOU KNOW?

Along with the variety of bird species comes a variety of names. Some birds have names that are brief descriptions of the bird's color or special features, such as the blue jay and the tufted titmouse. Other birds are named for their songs, such as the bobwhite and mockingbird; their habits and habitats, such as the marsh wren and woodpecker; their diets, such as kingfishers and flycatchers; or the person who discovered the bird, such as Cooper's hawk and Bewick's wren. This activity will help children become more aware of bird names and bird characteristics tied to those names.

## MATERIALS

- **BIRD PICTURES**
- **SMALL BROWN PAPER BAGS (ONE PER CHILD)**
- **NEWSPAPER**
- **CONSTRUCTION PAPER**
- **FEATHERS (LEGALLY OBTAINED)**
- **SEQUINS**
- **PIPE CLEANERS**
- **POM-POMS**
- **YARN**
- **AN ASSORTMENT OF COLLAGE MATERIALS**
- **SCISSORS**
- **GLUE**

## ACTIVITY

1. Display the various bird pictures, along with their names, in the art center.

2. Provide the remaining materials for children to create their own birds during self-selected activity time.

3. The body of the bird may be made by stuffing the paper bag with newspaper. Older children can do this themselves but younger children might need the bags prestuffed.

4. Feathers, construction paper, sequins, pipe cleaners, and other odds and ends can be used to complete the birds.

5. Once children have finished creating their birds, Ask them open-ended questions like these:
   - What does your bird eat?
   - What does your bird's song sound like?
   - What is your bird's habitat?
   - How does your bird's beak and feet help it in its habitat?

## IDEAS FROM SHERRI'S CLASSROOM

This learning experience was the result of the children wanting to create the birds they observed at the feeders. They brainstormed materials for creating the birds and spent several days putting their birds together. After hanging their birds in the classroom, they extended the experience to clay. Their representations of birds in a variety of media demonstrated their increased understanding of the various parts of birds and bird habits.

## ADDITIONAL ACTIVITIES

- **ART**—Have clay or any type of playdough available for children to sculpt birds during self-selected activity time. As they work, talk about a bird's body parts.
- **BULLETIN BOARD**—Hang the completed birds from the ceiling so they appear to be flying. In the fall or early winter, hang them so they are flying south. Change their direction to north as spring approaches.
- **PRETEND PLAY**—Make a pair of bird wings from a set of pillow cases by sewing a long narrow seam on one side of each pillowcase. Children can slide their arms into the space and pretend to be birds. Pillowcases could be painted to present more realistic wings.
- **STORY**—Read *Have You Seen Birds?* by Joanne Oppenheim (see appendix A on p. 180). This book provides an introduction to many species of birds and their habitats. Many of the pictures provide a "bird's eye" view of the environment.
- **WRITING**—Provide bird field guides for children to look through. During self-selected activity time, put out small blank books for them to name their birds, write descriptions of their birds, and describe their habits and habitats. Provide an audio recorder for children to record their bird's unique song.

# Explore a Feather 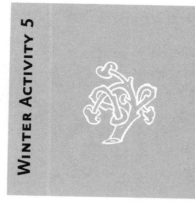 Science

## DID YOU KNOW?

Although not all species of birds can fly, all have feathers and wings. Bird feathers are tough, lightweight, and strong. The feathers are slightly oily, which allows them to shed water. Birds' feathers also keep them warm. This activity will enable children to explore feathers and discover their importance to birds.

## MATERIALS

- **FEATHERS (LEGALLY OBTAINED)**
- **PAN OF WATER**
- **MAGNIFYING GLASSES**
- **SCALES**
- **SCIENCE JOURNALS**

## ACTIVITY

1.  Place the feathers and other materials in the science area and have them available for children to explore during self-selected activity time.

2.  Allow children to explore the feathers—place in water, touch, examine with magnifying glasses, pull apart, weigh, run with them, and so on.

3.  Ask children open-ended questions like these:
    - How do you think feathers help birds when it rains?
    - Why do you think birds have feathers?
    - How would you describe a feather?

4.  Encourage children to document observations and responses in their science journals.

## Additional Activities

- **Bulletin board**—Take pictures of the children exploring the feathers. Display the pictures along with their thoughts, ideas, and dialogue about the bird feathers for parents to see and for children to revisit and reflect on their experience.
- **Group**—Discuss and document all the ways children think a feather can be used.
- **Story**—Read *My Feather* by Jane Mainwaring (see appendix A on p. 180). This nonfiction book explores basic feather concepts.

# I'm a Cardinal &#42; Music

## DID YOU KNOW?

Birds can be identified by color, profile, and call. Cardinals are a common bird throughout the Midwest, Southeast, and eastern United States. They are an easily identifiable bird found in hedgerows, wood margins, and suburbs. Their heavy conical beak is well adapted for cracking seeds. This activity will familiarize children with some of the unique characteristics of the cardinal.

## MATERIALS

- **PICTURE OF A CARDINAL**

## ACTIVITY

1. During group time, show children the cardinal picture. Discuss some of the distinguishing characteristics of the cardinal.

2. Teach children the following song to the tune of "Frère Jacques":

   I'm a cardinal, I am red. (*Fold arms like wings and flap them.*)
   I have a tuft, up on my head. (*Make a point on head with hands.*)
   I sing a pretty song (*Flap arms.*)
   and my beak is small and strong. (*Make beak with fingers.*)
   So I can crack seeds (*Pretend to peck at seeds.*)
   when I feed.

3. Once children have learned the song, ask open-ended questions like these:
   - How is the cardinal different from other birds? The same?
   - How do cardinals live in the winter?
   - How do cardinals' beaks and feet help them survive?
   - What other birds have you seen with a tuft?
   - What other birds have a beak like a cardinal?

## IDEAS FROM SHERRI'S CLASSROOM

This song is always a favorite in my classroom. The cardinal's readily identifiable physical characteristics make it easy for young children to spot and identify. The song helps them identify distinguishing characteristics of the cardinal. If cardinals aren't native to your region of the country, create a simple song about a common bird indigenous to your area.

## ADDITIONAL ACTIVITIES

- **ART**—During self-selected activity time, challenge children to record the songs of birds in ways other than by making an audio recording. For example, they might write, draw, color, and so on.

- **FIELD TRIP**—Select a special area to take children for a birdwatching walk. Before the field trip, explore the area yourself to see what birds are there. Using an audio recording of birdcalls, select two or three birdcalls for children to identify on the birdwatching walk. Have pictures of the birds available for children to associate the call with the visual images of the birds.

- **MUSIC**—Purchase or borrow from the public library an audio recording of birdcalls and play it during self-selected activity time.

- **OUTSIDE**—Sit quietly outside and listen to bird songs. As a result of spring mating, songs will be more prevalent from March through June.

- **PRETEND PLAY**—Provide binoculars for children to birdwatch in the pretend-play area during self-selected activity time. Cover the ends of two cardboard tubes with colored cellophane, tape them together, and attach a yarn strap.

- **SCIENCE**—Encourage children to watch for birds with conical beaks (such as finches, sparrows, grosbeaks, and buntings) at the birdfeeder. Look for similarities and differences in males and females. Provide science journals for children to document their thoughts, observations, and ideas.

- **STORY**—Read *Bird Talk* by Ann Jonas (see appendix A on p. 180). This picture book features common birds in their natural environment and uses memory phrases or words that naturalists often use to recognize and remember birdsongs.

# Migration Obstacles ❧ Large Motor

## DID YOU KNOW?

The seasonal movement of birds and other animals is called *migration*. Migration may be triggered by reduced hours of sunlight, a food shortage, or colder temperatures. There are many theories as to how birds find their way.

Some ornithologists believe some birds follow mountain ranges, coastlines, or rivers. Birds flying over the ocean are thought to follow the stars or angle of the sun. This activity will introduce the concept of bird migration to children.

## MATERIALS

- EQUIPMENT FOR AN OBSTACLE COURSE (SUCH AS A BALANCE BEAM, CLIMBER, CARPET SQUARES, MATS, STREAMERS, STUFFED ANIMALS, AND SO ON)

## ACTIVITY

1. Set up an obstacle course using any materials at hand. Keep in mind various migration scenes the course could represent, such as a balance beam "river," climber "mountain," streamers hung for "rain," carpet squares representing "food," large boxes for "buildings," stuffed animal "predators," and so on.

2. Before taking children through the course, discuss and define migration. Ask them open-ended questions like these:
   - Why do you think some birds migrate and others don't?
   - How do you think birds know where to go in the winter?
   - How do birds know where it is safe to rest and find food?
   - What dangers must birds look out for?

   Keep a record of the children's ideas about things a bird might experience during migration.

3. Move through the course, telling a migration story as you go. Allow the children to move through the course making up their own migration stories.

## ADDITIONAL ACTIVITIES

- **OUTSIDE**—Observe birds migrating.
- **SCIENCE**—Repeat the birdfeeder activity (see winter activity 1) during spring, summer, and fall. Compare the variety of birds seen at the feeder each season.
- **STORY**—Read *Goodbye Geese* by Nancy White Carlstrom (see appendix A on p. 179). In this book, a young child asks his father about the arrival of winter and migrating geese.

# Track Puzzlers ❧ Manipulative

## DID YOU KNOW?

Animals can be identified by their tracks. Winter and early spring are often good times for observing tracks in the wild due to weather and soil condi- tions. This activity will help children begin to associate animals with their tracks.

## MATERIALS

- **WINTER PATTERN 2 (SEE APPENDIX D, PP. 220–222)**
- **SET OF THE ANIMAL TRACK PUZZLES, LAMINATED AND CUT APART**

## ACTIVITY

1. Place the animal track puzzles on a table and allow children to play with them during self-selected activity time. Use only a few of the puzzles for younger children and more for older children.

2. Discuss the distinguishing characteristics of each animal's track as the children are playing. Ask them open-ended questions like these:
   - How can you tell which animal this track belongs to?
   - What animal tracks have you found in the snow or mud?
   - Do you think this animal has a foot or hoof? How many toes do you see?
   - Why do some tracks show claws and others don't?

## Additional Activities

- **Art**—During self-selected activity time, help children create child tracks by painting the bottom of each child's bare feet and having them step on a large sheet of paper.
- **Bulletin board**—Make a crayon rubbing of the bottom of each child's shoe. Create a display of these tracks, along with pictures of the children.
- **Field trip**—Visit a natural area to look for tracks. Other animal signs children might look for include chewed nuts or stripped cones; scratch marks, holes, or gnaws on trees; nests; and scat (animal feces). Be sure to take along science journals, magnifying glasses, and field guides.
- **Outside**—During self-selected activity time, have children carefully cover several cookie sheets with peanut butter. Before going outside, discuss with children where they think animals may live on the play yard. Decide where to place the cookie sheets to capture animal signs. Place the cookie sheets in the designated positions and check each day for animal signs. Discuss what animal made the various signs in the peanut butter.
- **Outside**—In a shallow bed of wet sand have children make tracks of their shoes. Mix children up by asking them to run around, then challenge them to match up their shoes with the tracks.
- **Outside**—Look for signs of animals while being careful not to disturb them. Be sure to look under birdfeeders.
- **Science**—Put out a track identification guide for children to explore during self-selected activity time.
- **Story**—Read *Animal Tracks* by Arthur Dorros (see appendix A on p. 179). This story introduces the tracks and signs left by various animals.

# Willie the Woodchuck ❧ Story

## DID YOU KNOW?

Animals hibernate in caves, hollow logs and trees, and holes in the ground. Many mammals move in and out of hibernation throughout the winter. In northern regions of the United States, woodchucks or groundhogs go into deep hibernation from November to February. They mostly eat plants but also bugs and snails. This activity will help children become aware of what hibernation is and how animals prepare for it.

## MATERIALS

- **WINTER PATTERN 3 (SEE APPENDIX D, PP. 223–224)**
- **FLANNELBOARD CHARACTERS**
- **FLANNELBOARD**
- **PICTURE OF A WOODCHUCK**

## ACTIVITY

1. Show the picture of the woodchuck and discuss his habits and habitat. Read the following story to children, placing the felt characters on the flannelboard at the appropriate times.

2. After reading the story, ask children open-ended questions like these:
   - How do you think Willie knows when to go to sleep and when to wake up?
   - What do other animals do in the winter?
   - Why doesn't Willie need to wake up to eat?

## ADDITIONAL ACTIVITIES

- **BLOCK**—During self-selected activity time, provide materials for children to build shelters for animals to hibernate in or find shelter during the winter.
- **MANIPULATIVE**—During self-selected activity time, provide animal pictures for children to sort animals by those who hibernate and those who don't.
- **NUTRITION**—During lunch or snack ask children to pretend they are animals preparing to hibernate. They need to eat a lot to get fat and grow a thick coat. At naptime, ask them to curl up on their cots and pretend to sleep like ani-

Willie was a woodchuck.
He had a bushy tail.
Every toe on each foot
had a long sharp nail.

He used his feet to dig a hole.
He dug deep down.
He dug it in a fence row,
where wheat grew all around.

Every day at sun up,
he'd search for plants and bugs.
When he was full each night, he'd
  sleep
in the hole that he had dug.

Willie soon got nice and fat
from wheat that he chomped down.
He'd lay for hours in the sun
by his hole in the ground.

Soon the air was cooler
and the leaves turned brown.
Then the fall winds came,
and the leaves fell down.

Winter was a-coming.
When Willie heard the news,
he crawled down in his woodchuck
  hole
to take a little snooze.

B-r-r-r came the winter cold.
Whoosh the wind would blow.
Willie kept on sleeping
and down came the snow.

Down came the winter snow.
Soon it was very deep.
Down inside his woodchuck hole
was Willie, fast asleep.

Down came ice and snow.
The sky was dark and gray.
Still Willie kept on sleeping
day after day.

The first spring warmth came.
The trees began to bud.
The ice and snow began to melt
and dirt turned into mud.

Down in Willie's woodchuck hole,
it was dark and deep.
Willie opened up his eyes
and woke up from his sleep.

He crawled out of his woodchuck hole.
He crawled outside to see.
Looking for those plants and bugs
as hungry as can be.

Soon he was back to searching
for seeds, plants, or bugs
and snoozing in the sun next to
the hole that he had dug.

In the months before the winter
Willie eats all he can hold
and goes to sleep all winter
so he doesn't feel the cold.

—John Griffin

mals in hibernation. When it's time to get up, tell them it's spring.

- **OUTSIDE**—During self-selected time, encourage children to do one of the following:
  - Find places on the play yard that are warmer or protected from the cold or wind.
  - Find good places to hibernate on the playground.
  - Look for dormant plants in the play yard.
- **PRETEND PLAY**—Place a box "cave" in an area where children can freely crawl in and out of it to pretend to hibernate during self-selected activity time. The cave could be painted by children to provide a more realistic appearance. Large rocks and pools of water might also be added.

- **STORY**—Read *Animals in Winter* by Henrietta Bancroft (see appendix A on p. 184). This book describes how various animals prepare for winter.

# Skaters Away ❋ Large Motor

### DID YOU KNOW?

People around the world enjoy a wide variety of winter sports. This activity will help children identify at least one way peo-ple use natural resources in the winter.

## MATERIALS

- **UNIT BLOCKS OR WAXED PAPER AND RUBBER BANDS**
- **CARPETED AREA**
- **MUSIC**
- **CHART PAPER AND MARKER**

## ACTIVITY

1. Talk with children about the kinds of things people do outside for fun in the winter. On chart paper, write their responses to open-ended questions like these:
   - What do you like to do outside in the winter?
   - What do you do in the winter that you can't do in the summer?

2. Give each child two unit blocks and explain that these are their ice skates and today they are going to go skating. Skates are manipulated by placing the feet on top of the blocks and sliding the block on the carpet. (Blocks can be waxed prior to the activity to make them glide easier, or waxed paper can be attached to feet with rubber banks if blocks are not available.)

3. Play "Skater's Waltz" or other music for skating and encourage children to glide through the room.

## Ideas from Sherri's Classroom

This activity was created by the children in my classroom after several visited a local ice rink. It allows the children to experience a winter sport while using large-muscle skills.

## Additional Activities

- **Art**—During self-selected activity time, choose from one of the following:
  - Provide magazines for children to cut out pictures of people doing various winter activities. Glue these pictures onto paper to create winter collages.
  - Provide materials for children to create "snow" sculptures from Styrofoam.
  - Create snow globes by gluing objects to the lid of a small, clear jar. Add water and glitter. Seal and shake.
  - Have the children form an ice sculpture. Ask parents to freeze containers of water (unless you have enough freezer space). On the day the ice sculpture is to be built, put all of the ice in a small child's swimming pool. Show children how to attach ice pieces together by sprinkling salt on the ice and spraying with water. Provide spray bottles of colored water to decorate the sculpture when it's finished.
- **Block**—Use the fish from "Let's Go Fishing" (see summer activity 12). Place them inside a box with a lid. Cover the lid with white paper and cut a large hole. During self-selected activity time, use the poles and go "ice fishing."
- **Block**—Build snowmobiles with large hollow blocks and make goggles out of elastic and plastic six-pack rings.
- **Bulletin board**—Photograph children participating in various winter activities and display the photos along with their ideas recorded during this activity.

- **Group**—Chart or survey children's preferences for the various winter activities.
- **Large motor**—Encourage children to organize and set up their own "Winter Olympics."
- **Large motor**—Put on slow, calm music and encourage children to pretend to be melting snow people or sculptures. Be sure to discuss why the snow people or sculptures melt.
- **Music**—Sing "This Is the Way We Shovel the Snow" to the tune of "Here We Go 'Round the Mulberry Bush." Encourage children to make up other verses about things they do in the snow.
- **Nutrition**—Provide materials for children to create "snow" sculptures for snack with large and small marshmallows, raisins, pretzels, and so on.
- **Outside**—During outside self-selected activity time, put out small snow shovels for children to try shoveling snow.
- **Pretend play**—Provide small carpet squares for children to use as sleds.
- **Story**—Read *The First Snowfall* by Anne Rockwell and Harlow Rockwell (see appendix A on p. 189). This story illustrates fun things to do in the snow.

# What's the Temperature? ❧ Science

## DID YOU KNOW?

Outside temperature is a major weather indicator. This activity will help children become more aware of outside temperature and how the temperature affects them.

## MATERIALS

- TWO THERMOMETERS (ONE FOR OUTSIDE AND ONE FOR INSIDE)
- STRIP OF ¼- TO ½-INCH ELASTIC
- STURDY PIECE OF CARDBOARD
- PICTURES OF CHILDREN IN CLOTHING FOR DIFFERENT TEMPERATURES
- GLUE
- RED MARKER

## ACTIVITY

1. Prepare a cardboard thermometer by drawing a thermometer on the cardboard. Color half the elastic strip red. The red will be the colored alcohol found in most thermometers. Put the elastic through slits in the top and the bottom of the "thermometer." Sew or pin the ends together in the back. This will enable the "alcohol" to be moved up and down.

2. Glue the pictures of children in various clothing at the appropriate places on the cardboard thermometer.

3. Place the two real thermometers inside and outside in places where children will notice, read, and study them. Place the cardboard thermometer near the real ones. As children discover the thermometers, challenge them to compare and manipulate the cardboard thermometer. Ask children open-ended questions like these:

- How do you know what to wear when you want to go outside and play?
- How do you think the temperature affects animals?
- What causes the temperature to change?

4. As children show interest in the thermometers, discuss the thermometers' purpose. Compare observing the outdoor's temperature to taking the children's temperatures. Encourage children to compare the outside thermometer with the inside one.

## ADDITIONAL ACTIVITIES

- **BULLETIN BOARD**—Display pictures of winter scenes or pictures of animals in winter environments.
- **GROUP**—Record daily temperature and general weather conditions on a chart.
- **GROUP**—Create a chart that illustrates the number of children who wore hoods, stocking caps, or baseball hats on any given day.
- **MANIPULATIVE**—During self-selected activity time, provide pictures of children in different clothing for sorting. Discuss why children are wearing the various articles of clothing.
- **OUTSIDE**—Encourage children to notice the activity patterns of animals as they read the thermometer. The birdfeeder is a good place to observe animal activity levels.

- **PRETEND PLAY**—Set up a weather station in the pretend-play area. Include maps, weather forecaster dress-up clothes, pointers, thermometers, and so on.
- **PRETEND PLAY**—Put out seasonal clothing in the pretend-play area.

# What Is Energy? ❧ Group

## DID YOU KNOW?

Most everything that moves requires energy. This activity will encourage children to think about energy and its sources.

## MATERIALS

- **CHART PAPER AND MARKER**

## ACTIVITY

1. During group time, discuss with children what they know about energy. On chart paper, write their responses to open-ended questions like these:
   - What is energy?
   - Where does energy come from?
   - How do you know when you are using energy?
   - What needs energy?

2. Play a game such as "Duck-Duck-Goose" or "Ring Around the Rosy," and discuss how the children feel after playing the game.

3. Talk about the energy they used to play the game and where they got that energy.

## ADDITIONAL ACTIVITIES

- **FIELD TRIP**—Visit places that provide energy (such as a power plant, service station, or grocery store).
- **NUTRITION**—Discuss children refuelling their own energy as they eat snack or lunch.

Energy use is an abstract concept for young children to grasp. However, attitudes about and habits of energy consumption are formed at a young age. Relating energy use with children's own bodies makes energy consumption more personally meaningful thus allowing children to consider and become aware of other types of energy use.

- **SCIENCE**—In small groups, have children stand where the sun shines brightly into the classroom, and then compare how they feel when they stand where the sun isn't shining.
- **SCIENCE**—Show children how to experiment with static electricity by rubbing a balloon on a child's hair and sticking the balloon to the wall. Rub some wool on a comb and show how the comb will pick up small pieces of paper. Provide materials for children to experiment on their own.
- **STORY**—Read *Energy Makes Things Happen* by Kimberly Brubaker Bradley (see appendix A on p. 181). This nonfiction book clearly defines energy, where it comes from, and how it is transferred from one thing to another.
- **WOODWORKING**—Provide nonworking appliances, with the cords removed, for children to disassemble during self-selected activity time. Talk about how the appliance once used energy to perform tasks. Encourage children to try to figure out how the appliances worked as they take them apart.

# B, T, and U  Story

### DID YOU KNOW?

The sun provides much of the natural energy used on our planet. This story will help children develop an understanding that energy comes from the sun as well as ways people use various forms of energy.

## MATERIALS

- **WINTER PATTERN 4 (SEE APPENDIX D, PP. 225–226)**
- **FLANNELBOARD**
- **FELT STORY PIECES**

## ACTIVITY

1. Read the following flannelboard story to children, placing the felt pieces on the flannelboard at the appropriate times.

2. After reading the story, ask children open-ended questions like these:
   - What are some other kinds of energy *B*, *T*, and *U* might become?
   - What are some things you could do with paper grocery bags?
   - What are some ways that *B*, *T*, and *U* are wasted?

Way high in the sky
important things are done.
Little bits of energy
are made inside the sun.

Down from the burning sun
warm pieces flew.
Three little pieces
named B, T, and U.

From the sun to our earth
the little pieces hurried.
"Will they use us wisely?"
they wondered and they worried.

B hit a plant leaf
at the end of his fall.
Zap! the plant leaf used him
to grow a cotton ball.

A farmer picked the cotton
and spun it into cloth
and sewed it all into a bag
very large and soft.

A cloth bag you can use
again and again.
A shopper bought the cloth bag
to carry groceries in.

Those brown paper grocery sacks
are all made out of wood.
So, use cloth bags and save more
   trees
like everybody should.

T was the next piece
he dropped into the sea
and heated up the water there
as warm as it could be.

Invisible, so we can't see
he made the water rise.
Up into a cloud it went
floating in the sky.

Down came the water
in sheets of pouring rain.
Whoosh! down a river
like a roaring train.

A dam held the river
on its way back to the sea,
and slowing down the water
made some electricity.

Zap! Through a power line,
T traveled on and on,
and into a light socket
right there in your home.

Voom! He makes the sweeper go,
and the TV too.
Are there many other things
that he can do?

Switch! He makes the lights come
   on.
But how long will he last?
What are some things that we can do
so he won't get used so fast?

U was the last piece.
Down from the sun he flew.
Right into a field
where a corn plant grew.

Changed into the yellow corn
in a shuck of green.
He waited for the farmer
and his corn-picker machine.

The grocer bought the piece of corn
and put it on a rack.
A shopper took it home with her
in a cloth shopping sack.

She turned on her electric stove
to make the burner warm.
Then she boiled some water
to cook the piece of corn.

A small child ate the corn
and as he chewed he looked
at all the things that made the corn
that his mother cooked.

B made the cloth bag
to bring home things to eat.
T made electricity
to cook the corn with heat.

And U made the piece of corn
the child ate that day.
So he would have the energy
to live and learn and play.

—John Griffin

## Additional Activities

- **Art**—Provide catalogs and magazines for children to find pictures of things that use energy, cut them out, and create collages during self-selected activity time.
- **Block**—Create small gas tanks out of cardboard tubes from toilet paper rolls or milk jugs to use with cars and trucks during self-selected activity time.
- **Music**—Sing "Row, Row, Row Your Boat" and discuss the type of energy a rowboat uses.
- **Nutrition**—Make decaffeinated sun tea with the children. Talk about how the sun helped the process.

- **Outside**—Make gas tanks from large cardboard boxes with a hose attached. Place them outside for use with the tricycles.
- **Science**—Place a plant in the sun and another in a dark room. Encourage children to note the differences between the plants after a few days and again after a week. Provide science journals for children to document their observations.
- **Science**—Bring in a selection of windup toys for the science area during self-selected activity time. Discuss where they get their energy.
- **Story**—Read *Alexander and the Wind-Up Mouse* by Leo Lionni (see appendix A on p. 181). This is the story of the relationship between a real mouse and a windup mouse. It provides the basis for discussion about the two types of energy they use.
- **Story**—After reading the story at group time, have the flannelboard and story pieces available in a learning center to use during self-selected time.

# No Electricity 🐢 Group

## DID YOU KNOW?

Conservation means wise use of natural resources. People need to practice conservation so there will be natural resources left for the future. This activity will help children discover what life is like without electricity. Conservation attitudes formed now will remain with children as they mature.

## MATERIALS

- **KEROSENE LAMPS OR CANDLES**
- **CLEAN BUCKET OF WATER**
- **DIPPER**
- **CUPS**

> This activity requires a dark room; you can close the blinds or wait for a cloudy day.

## ACTIVITY

1. Experience a day without electricity.

2. Put out the kerosene lamps or candles and the bucket of water with cups for drinking. Be sure to place the lamps and candles out of children's reach and caution children about the flames.

3. As children go about their day, discuss how the day is changed without electricity. Talk about what life would be like without electricity. Ask children open-ended questions like these:
   - How can we keep warm without electricity?
   - How can we keep food cold or cook food without electricity?
   - What can you do for fun when there isn't any electricity?
   - What would happen if we ran out of electricity?
   - How can we keep from wasting electricity?

## ADDITIONAL ACTIVITIES

- **FIELD TRIP**—Visit a store where woodstoves are sold and talk about whether or not wood stoves are a wise use of our natural resources. Renewable and nonrenewable resources could be discussed. Oil and coal are nonrenewable while wood is a renewable resource—that is, trees can be replanted.

- **GROUP**—Discuss other natural resources and how children can conserve or use them wisely.

# Does the Sun Give Us Energy? ❀ Science

## DID YOU KNOW?

The sun gives us free energy every day. This activity will help children become more aware of the sun's energy.

## MATERIALS

- **THREE PIECES OF DARK CONSTRUCTION PAPER**
- **BOOK**
- **FLOWER POT**
- **SCIENCE JOURNALS**

## ACTIVITY

1. During self-selected activity time, place the three pieces of paper in a sunny window.

2. Leave the first piece uncovered. Place a book on the second piece so that only part of the construction paper is covered. The flower pot should be placed in the center of the third piece of construction paper.

3. Ask children to predict what they think will happen to the construction paper. Encourage them to document their ideas in their science journals or by making a poster to show their predictions and the actual results.

4. Ask children open-ended questions like these:
   - What causes you to get a sunburn?
   - How do you think the sun makes such a powerful light?
   - Find some objects to place on the paper that you think the sunlight might pass through.
   - Where is a place in our room that is heated by the sun?

5. Remove the book and flower pot after a few days. Compare children's observations and their predictions, and document the results.

## Ideas from Sherri's Classroom

This activity is based upon a discovery made by the children one day when we were cleaning our classroom. Construction paper had been used as matting for pictures and artwork placed on a bulletin board near a sunny window. When the work was taken down, the children noticed the discoloration of the construction paper that wasn't covered by a picture or art work. This led to a more focused experiment with the children intentionally placing various objects on the construction paper and deliberately placing the paper in the sun.

## ADDITIONAL ACTIVITIES

- **ART**—Provide various materials for children to create sun pictures during self-selected activity time. Discuss life without the sun and explain that the earth's only source of heat is the sun.

- **ART**—Cut out advertisements using warm and cool colors (for example, soup advertisements often use reds and oranges to make you feel warm while toothpaste advertisements use blues and greens to make you feel cool). Discuss with children the use of warm and cool colors and how they make people feel. Sort advertisements into warm and cool feelings. Use of warm and cool colors is an important component to advertising. Compare the warm and cool colors with what happened in the experiment. Provide old magazines for children to create warm and cool collages. Notice colors used to designate hot and cold water on many faucets.

- **BULLETIN BOARD**—Display pictures of people using heat from the sun (such as clothes hanging outside, crops growing, and so on).

- **GROUP**—Discuss how some animals' coats are darker in winter than in summer (for example, deer, coyote, and dogs).

- **NUTRITION**—Let children try drying various fruits in the sun (such as grapes or apples).

- **OUTSIDE**—Fill spray bottles with heavily colored water and allow children to paint the snow during outside self-selected activity time. Discuss changes in the snow and if one color of water causes the snow to melt faster than another color.

- **OUTSIDE**—On a sunny day, feel the temperature difference between a white car and a black or red car.

- **SCIENCE**—Set up a "control" experiment by duplicating the experiment but put one set of papers in the dark or a place where the sun doesn't shine.

- **SCIENCE**—Try other sun experiments, such as leaving crayons outside in the sun. Encourage children to brainstorm ways to find out more about the sun.

- **STORY**—Read *Who Likes the Sun?* by Beatrice Schenk DeRegniers (see appendix A on p. 181). This book tells a story of a child enjoying the sun.

# Saving Energy ❧ Group

## DID YOU KNOW?

People should practice conservation or wise use of natural resources so they can be used by more people over a longer period of time. This activity will help demonstrate some conservation or wise use practices for children.

## MATERIALS

• **THIS ACTIVITY REQUIRES NO SPECIAL MATERIALS**

## ACTIVITY

1. Gather children at group time and tell them you are going to play an energy-saving game. You will describe a way that energy is being wasted and they can tell you some ways to practice conservation or wise use.

2. Describe the following situations to the children and encourage them to share their ideas about ways to prevent the energy from being wasted.
   • Chelsey goes into the bathroom to wash her hands. Her mother calls her to dinner and Chelsey leaves the faucet running.
   • Zachary goes into the living room to watch television. He soon becomes interested in playing with Lego building blocks and leaves the room to look for more pieces, but he forgets to turn off the television.
   • Austin drinks a bottle of juice on his way home from school and throws the plastic bottle on the ground.

   • Jacob likes to draw. His mother bought him a new tablet. Jacob makes a few marks on each page and then throws the tablet away and asks his mother for a new one.
   • Ashlyn misses her school bus, so she and her mother jump into the car and drive fast to school.
   • Eleanor wants to put some birdseed on the ground outside her window so she can watch the birds eat. It's winter and the birds need extra food because the snow on the ground is covering up all their food. Eleanor opens the window and puts the birdseed out but doesn't close the window.

3. Ask children open-ended questions like these:
   • What are some ways that we waste energy at school?
   • What would happen if we used up all of our natural resources?
   • What can we do to make sure we don't waste things?

## ADDITIONAL ACTIVITIES

• **GROUP**—Relate the term *conservation* to wildlife and plants. Discuss ways that people use natural resources.

# Grocery Shopping ✿ Pretend Play

## DID YOU KNOW?

One way to reduce our use of natural resources is to reuse materials. In recent years, grocery stores have started providing cloth bags with their logos for shoppers to reuse rather than always using paper or plastic bags. This activity will help children make a connection between reusing natural resources and something that occurs in their daily lives, and relate the concept of wise use to real-life experiences.

## MATERIALS

- **SEVERAL CLOTH SHOPPING BAGS**
- **EMPTY FOOD CONTAINERS (SUCH AS CEREAL BOXES, EGG CARTONS, CANNED GOODS)**
- **CASH REGISTER (THIS CAN BE MADE FROM A BOX)**
- **TELEPHONE**
- **PLAY MONEY**
- **PAPER**
- **PENCILS**
- **RECEIPT BOOK**

## ACTIVITY

1. Set up a grocery store in one corner of the classroom for children to use during self-selected activity time.

2. Price grocery items. Put play money in purses and billfolds. Place the cash register and telephone near the checkout area.

3. Be sure to place paper and pencils near the phone for messages. Receipt books can be used to provide receipts with purchases. Additional writing materials should be available for making grocery lists.

4. Encourage children to pack their purchases in cloth shopping bags.

5. Discuss the various food packaging materials and what happens to it once the food is eaten. Ask children open-ended questions like these:
   - What would happen if everyone used cloth shopping bags at the grocery store?
   - What if no one used cloth shopping bags?
   - How could we encourage people to use fewer paper or plastic bags?
   - Where do you think the food in the grocery store comes from?
   - What happens to the package after we eat the food?

## ADDITIONAL ACTIVITIES

- **ART**—As you use cardboard tubes from toilet paper rolls, egg cartons, Styrofoam packaging, and other materials in the art area, talk about the original purpose of each.
- **FIELD TRIP**—Visit a grocery store and examine various types of product packaging.
- **GROUP**—Discuss strategies that children might use to help their parents remember to use their cloth shopping bags.

# Fred's Forest ❀ Story

## DID YOU KNOW?

Conservation means the wise use of natural resources. There are three levels of conservation: preservation, restoration, and management. This story will introduce the three levels of conservation to children.

## MATERIALS

- **WINTER PATTERN 5** (SEE APPENDIX D, PP. 227–228)
- **FLANNELBOARD**
- **FELT CHARACTERS**

## ACTIVITY

1. Read the following poem to children, placing the felt characters on the flannelboard at the appropriate time.

2. After reading the poem, ask children open-ended questions like these:
   - What would have happened if Fred hadn't planted more trees?
   - How could Fred make his firewood last longer?
   - How do you heat your house?

## ADDITIONAL ACTIVITIES

- **BLOCK**—Create trees from twigs and clay for children to use in the block area during self-selected activity time.
- **FIELD TRIP**—Visit a forest.
- **GROUP**—Discuss other forest products such as nuts, lumber, paper, and so on.
- **STORY**—Provide felt characters for children to retell the story during self-selected activity time.

## Ideas from Sherri's Classroom

Trees are one of North America's most valuable renewable natural resources. Properly managed, forests can be productive sources for firewood and other forest products as well as wildlife habitat. Forest management can even be directed to favor certain species of wildlife while still providing products. For example, some old-growth forests house certain rare and endangered species. Clear-cutting or burning small areas of the forest will encourage regeneration or growth of young trees and brush that will benefit other species. Selective harvest of trees can increase the health of a forest by eliminating disease or insect infestation and opening up small areas of the forest for regeneration of trees.

Fred had a forest,
a home for birds and bees.
Oftentimes, from his porch
he'd sit and watch the trees.

Cut a tree? "No!" said Fred,
"They're nice to have around.
I like to sit and watch them.
I'll never cut one down."

But winter cold soon changed his mind,
with it's snow and sleet.
"B-r-r-r!" said Fred, "I need a fire.
I must have some heat."

Fred got a chainsaw.
He sawed some trees for wood.
He burned them in his fireplace
and the heat felt good.

Fred said, "I like this heat.
No more cold for me."
Fred got his chainsaw
and cut down every tree.

He stacked up stacks of firewood
that were higher than his head.
"I may not have a forest
but I'll be warm," he said.

The years went by, Fred's firewood piles
got smaller every day.
The ground where once the forest stood
began to wash away.

The animals all moved away.
Their homes were all cut down.
Fred wished he had his forest back.
It was nice to have around.

Fred then bought some baby trees.
He planted carefully,
but Fred soon discovered
it takes years to grow a tree.

Fred grew old and years went by,
winters, summers, fall.

Fred became an old man
before his trees grew tall.

Once again he had a forest.
The animals had a home.
And Fred was very, very proud
of the forest he had grown.

When Fred needed firewood
for heat in wintertime
he'd cut only trees he needed
and the forest did just fine.

The ground no longer washed away.
Fred's forest grew and grew.
When the trees got crowded
Fred would cut down one or two.

Fred's house was warm in wintertime.
He was warm as he could be.
Fred had lots of firewood
*and* a forest full of trees.

—John Griffin

# Insulation ✾ Science

## DID YOU KNOW?

Insulation helps keep heat in and cold out or cold in and heat out. This activity will help children become aware of insulation and its importance in conserving energy.

## MATERIALS

- **SEVERAL PAPER CUPS**
- **FOIL**
- **COTTON**
- **COTTON CLOTH**
- **WOOL FABRIC**
- **NEWSPAPER**
- **CARDBOARD**
- **RUBBER BANDS**
- **CHART PAPER AND MARKER**
- **SNOW OR ICE CUBES**

## ACTIVITY

1. During self-selected activity time, place several cups of snow or ice in the science area. Ask: "Which cup of snow or ice do you think will melt first? Why?"

2. Challenge children to use the materials provided to keep the snow or ice from melting. Ask: "What other things could we use to insulate the cups?"

3. List the various choices for insulation on chart paper, and document the melting process every five minutes.

4. Talk with children about why they think one type of insulation worked better than another.

## ADDITIONAL ACTIVITIES

- **BLOCK**—Put out cotton in the block area for children to insulate their buildings.
- **FIELD TRIP**—Visit a new building being built. Observe insulation being installed.
- **GROUP**—Discuss ways people keep cold air out of their homes—insulation, storm windows, caulking, and so on. Ask children to think of some ways they keep cold air out of their homes. Look for ways cold air is kept out of the school.
- **SCIENCE**—Display insulation and caulking in the science area for children to examine. Remember to include science journals and magnifying glasses.

# Winter Fabrics ❧ Science

## DID YOU KNOW?

Different fabrics have different insulating qualities. This activity will help children identify some of the fabrics from which winter clothes are made.

## MATERIALS

- SEVERAL DIFFERENT WINTER FABRICS (SUCH AS WOOL, FLANNEL, CORDUROY, FUR, VELOUR, AND SO ON—AT LEAST TWO OF EACH)
- BOX WITH A LID
- SCISSORS
- CARDBOARD
- MARKER

## ACTIVITY

1. Create a "feely box" by cutting a hole about the size of a child's hand in one end of the box.

2. Place one set of fabric swatches in the feely box. Attach the second set of fabric swatches to the piece of cardboard, and write the name of the fabric underneath the swatch.

3. Put the feely box and the fabric-matching card in the science area for children to experiment with during self-selected activity time. Encourage children to match the pieces of fabric by touch alone. When observing children's experimentation, ask them open-ended questions like these:
   - Which fabric feels the warmest?
   - Who is wearing something that feels like this?
   - Why do you wear this fabric at this time of year?
   - What would happen if you wore it in the summer?

4. Challenge children to find people in the classroom wearing winter fabrics like those included in the feely box.

## ADDITIONAL ACTIVITIES

- ART—During self-selected activity time, provide various winter-fabric scraps, glue, scissors, and paper for children to make fabric collages.
- ART—Look through summer and winter catalogs and discuss the difference in clothing. During self-selected activity time, provide scissors, glue, paper, and catalogs or magazines for children to make a winter-clothing collage.
- BULLETIN BOARD—Display pictures of people in winter clothing and summer clothing.
- GROUP—Discuss the differences between animals' coats in the summer and the winter.
- PRETEND PLAY—Provide dark winter clothing in the dress-up area. Talk about why people wear dark colors in the winter.

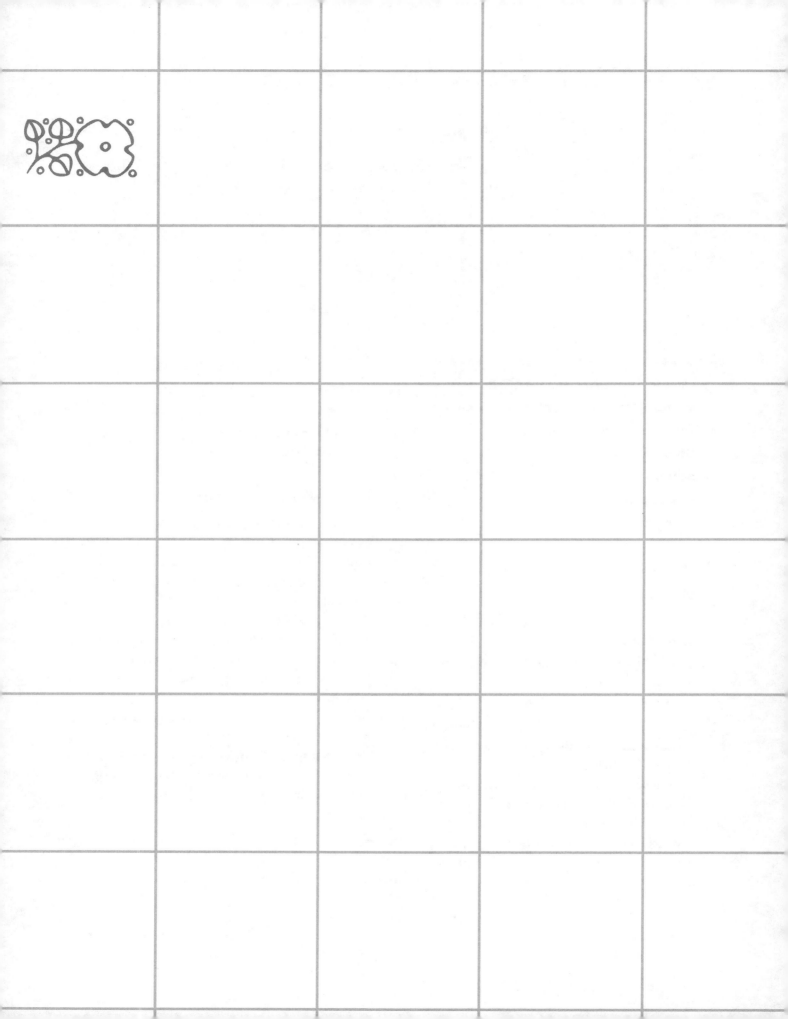

# Spring

Spring brings increased activity for children in the outdoors. Anticipation of warmer weather and longer days will entice you to plan outdoor activities.

Snow and ice begin to melt as a result of more direct radiation from the sun. Some insects awake to fly on tattered wings while others metamorphose from chrysalises or eggs. Migratory birds move to their nesting grounds and begin courtship behavior and nest building. Songbirds lay eggs, young squirrels begin to appear in trees, and larger animals begin to have their young. Tadpoles can be found in creeks, ponds, or temporary pools where toads and frogs lay eggs.

Planting season begins and many plants and trees begin to bloom. Wildflowers blooming from desert to wetland make the entire country a flowering wonderland. On the West Coast of the United States, humpback whales begin their northerly migration. Odds are greatest this time of year for spotting bears. As they emerge from hibernation, they are hungry and can be seen digging for roots or feeding on winter-killed mammals.

Spring holds several celebrations, including National Wildlife Week, Arbor Day, and Earth Day, and the spring equinox occurs with equal periods of darkness and light. These national observances give us cause to celebrate nature with young children.

Topics for spring activities in *My Big World of Wonder* include the following:

- Weather and seasonal changes
- Water cycle
- Clouds
- Wind
- Soil and planting
- Erosion
- Insects
- Birds
- Mammals

# Drip and Drop  Story

## DID YOU KNOW?

Water is constantly evaporating from the earth. It goes into the sky to form clouds. Eventually, it returns to the earth as precipitation. This story will introduce children to the water cycle, also called the hydrologic cycle.

## MATERIALS

- **SPRING PATTERN 1 (SEE APPENDIX D, PP. 229–230)**
- **FLANNELBOARD**
- **FELT FLANNELBOARD CHARACTERS**

## ACTIVITY

1. Read the following story (p. 98) to children, placing the felt characters on the flannelboard at the appropriate times.

2. After reading the story, ask children open-ended questions like these:
   - What kind of cloud do you think Drip and Drop lived in?
   - What causes thunder? Lightning? Fog? Snow? Hail?
   - Where did the water in the clouds come from?
   - Where do rainbows come from?

## ADDITIONAL ACTIVITIES

- **ART**—During self-selected activity time, choose from one of the following:
  - Provide materials for children to create rainbows using craypas, tissue paper, watercolors, colored yarn, and construction paper.
  - Provide children with construction paper, markers, crayons, paint, colored pencils, glue, and cotton balls. Encourage them to create cloud pictures. Cotton may be used to add texture to the clouds.
  - Put out materials for children to fingerpaint with dark-blue paint. As the children paint, talk about storms and how they make them

feel. After finishing their paintings, provide yellow yarn "lightning" for children to stick in the paint.

- **BULLETIN BOARD**—Photograph the various clouds children observe while outside. Record and transcribe their ideas about the clouds. Display the photographs, along with their discussions, so children and parents can revisit the experience.

- **FIELD TRIP**—Take children for a walk in the rain. Talk about how rain smells, feels, sounds, and tastes. Discuss how things change colors when they get wet. Document the field trip by taking photographs and making an audio recording of the rain.

- **LARGE MOTOR**—Create rainbow streamers by attaching strips of colored crepe paper to empty paper towel rolls. Play soft music and encourage children to make their rainbows dance in the rain.

- **MUSIC**—Teach children the following song to the tune of "Three Blind Mice":

  Three rain drops. (*Hold up three fingers.*)
  Three rain drops.
  See how they fall. (*Make fingers fall like raindrops.*)
  See how they fall.
  They all fall down from the clouds in the sky.
  They water the plants so they won't die.
  The plants feed the animals who live nearby.
  Three rain drops. (*Hold up three fingers.*)

- **MUSIC**—Provide children with various rhythm instruments and encourage them to create rain rhythms, music, and dances. Make a videotape or audiotape and photograph their work to share with parents.

- **MUSIC**—Create audio recordings of rain falling on different surfaces, such as a tin roof, sidewalk, or grassy field. Talk about the various differences in sounds.

- **MUSIC**—Provide children with a Chilean rainstick to experiment with during self-selected activity time.

- **OUTSIDE**—During self-selected activity time, encourage children to lay on their backs and observe shapes, sizes, and movements of clouds.

- **OUTSIDE**—Place a rain gauge outside for children to measure the rain. One can be made from a tall, clear jar with straight sides and measurements marked with nail polish. Put out two gauges, one under a tree and one in the open. Compare the contents after a hard rain. If you're in a region where it snows, use it during the winter to measure snowfall.

Drip and Drop were water.
Their cloud was puffy white.
They lived there high above the
　　ground,
through morning, noon, and night.

The cloud was filled with drips and
　　drops.
They floated in the air.
Thousands, millions, drips and
　　drops,
were floating 'round up there.

As time went by more drips and
　　drops
began to join the crowd.
It wasn't long before their home
became a thundercloud.

"Boom!" went the thunder!
A loud and scary crash!
Down came the drips and drops.
"Zap!" the lightning flashed.

Drip hit an umbrella,
and splashed on someone's nose.
Drip she fell and landed
on someone else's toes.

Drip slipped onto the pavement
and down a little hill,
right into a puddle
that soon began to fill.

Down, down, wheee!
Down came Drop.
He landed on a window pane.
He landed with a plop!

Off onto a plant leaf.
From leaf to leaf he dropped.
And on a very big leaf,
that is where he stopped.

Soon the sun began to shine.
It shone down everywhere.
And heated little Drip and Drop,
back up into the air.

Soon their cloud was full again.
The rain fell soft and slow.
And as the sun shone through the
　　drops,
it made a large rainbow.

—John Griffin

- **PRETEND PLAY**—During self-selected activity time, put out raincoats, hats, umbrellas, and boots in the dress-up area.

- **SCIENCE**—During self-selected activity time, choose from one of the following:
  - Place prisms near a window for children to discover how to create rainbows in the classroom. Add flashlights and science journals for children to experiment and record their observations.
  - Let children create a cloud by putting several inches of hot water into a cold jar. Put the lid on and place an ice cube on the lid. As the warm air rises, it is cooled and forms fog, or a little cloud.
  - Each day, make a drawing of the clouds. Date each of these and keep charts of the weather so children can make comparisons between types of clouds and weather.
- **STORY**—Read *The Cloud Book* by Tomie dePaola (see appendix A on p. 189). This nonfiction book introduces the most common types of clouds and information about them.

# Hanging Out ❀ Pretend Play

## DID YOU KNOW?

Water is always evaporating from the earth. This activity will encourage children to con-sider this phenomenon and discuss the process.

## MATERIALS

- **DOLL CLOTHES OR PRETEND-PLAY CLOTHES**
- **TUB OF SOAPY WATER**
- **TUB OF CLEAR WATER**
- **CLOTHESLINE**
- **CLOTHESPINS**

## ACTIVITY

1. Set up a clothes-washing station in the pretend-play area or outside.

2. String the clothesline outside, low enough that children will be able to reach it to hang items.

3. During self-selected activity time, encourage children to wash and rinse the clothing, then hang it outside.

4. Ask open-ended questions like these to begin a discussion of what children see happening:
   - What do you think happened to the water?
   - Where could we find more information about this?

## ADDITIONAL ACTIVITIES

- **OUTSIDE**—During outside self-selected time, provide housepaint brushes and buckets of water. Encourage children to waterpaint in both sunny and shaded areas. Observe and discuss what happens.

- **SCIENCE**—Encourage children to observe a puddle or make a puddle on a plate. Measure and record the puddle's size. Use photographs to doc-ument the changes in the puddle and display along with children's ideas about the puddle. Be sure to have science journals available for chil-dren to document their ideas.

# Where Do Animals Go When It Rains? ❀❀ Manipulative

## DID YOU KNOW?

Animals need protection during heavy rains. They hide in caves, under trees or other vegetation, in hollow trees or logs, and so on. This activity will encourage children to think about where and when animals find protection.

## MATERIALS

- **SPRING PATTERN 2 (SEE APPENDIX D, P. 231)**
- **LAMINATED OR FELT PICTURES OF ANIMALS**
- **LAMINATED OR FELT PICTURES OF A CAVE, HOLLOW TREE, MUSHROOM, ROCK, AND LEAF**
- **FLANNELBOARD OR TABLE**

## ACTIVITY

1. During self-selected activity time, place animal and shelter pictures on a table in the manipulative area or near the flannelboard.

2. Encourage children to match up animals with places where they might find cover when it rains.

3. Be sure to include animals—such as fish, deer, or turtles—that challenge children to think. Ask them open-ended questions like these:
   - Where do you go when it rains? Why?
   - Where do you think these animals go when it rains?
   - How do the animals know it is a safe place?

## IDEAS FROM SHERRI'S CLASSROOM

This learning experience invariably results in discussions about animals seeking shelter differently than people. Children's thinking is challenged when fish are included. They are pushed beyond their egocentrism to consider the perspective of the animal. They talk about what it is like under water when it rains. They also talk about different kinds of rain. Just as they play outside when it is raining softly, many animals ignore a gentle rain. Children from rural areas also discuss what livestock do when it rains. The discussion always leads to children being more observant and sharing personal stories.

## ADDITIONAL ACTIVITIES

- **BLOCKS**—During self-selected activity time, put out twigs, branches, and plastic animals in the block area. Encourage children to find places for the animals to hide in the rain.
- **NUTRITION**—Serve animal crackers and water for snack. Discuss where the various animals might spend a rainy day.
- **STORY**—Read *Where Does the Butterfly Go When It Rains?* by May Garelick (see appendix A on p. 189). This story explores where various animals go to get out of the rain.

# What Goes in the Wind?  Art

## DID YOU KNOW?

People use the wind in many different ways. This activity will challenge children to create something to use in the wind.

## MATERIALS

- **VARIOUS KINDS OF PAPER, SUCH AS COFFEE FILTERS, PAPER NAPKINS, TYPING PAPER, CREPE PAPER, AND TISSUE PAPER**
- **STRING**
- **TAPE**
- **HOLE PUNCHES**
- **STAPLER**
- **PAPER CLIPS**
- **MARKERS**
- **GLUE**
- **GLUE STICKS**
- **STYROFOAM PIECES**
- **STRAWS**
- **TOOTHPICKS**
- **CRAFT STICKS**

## ACTIVITY

1. Place materials in the art area during self-selected activity time.

2. Challenge children to create something that will move or travel in the wind. Ask them open-ended questions like these:
   - What causes the wind?
   - How do people use the wind?
   - How is the wind destructive?
   - How does the wind help people? Plants? Animals?
   - How will your creation move in the wind?

3. As children finish their work, encourage them to try out their creations in the wind and then make necessary adjustments.

## ADDITIONAL ACTIVITIES

- **ART**—During self-selected activity time, choose from one of the following:
  - Encourage children to create wind chimes by tying found items (such as sticks, rocks, shells, and so on) to a base such as plastic rings from a six-pack, a wire hanger, or a branch. Listen to them in the wind.
  - Provide children with thin paint, straws, and smooth paper. Encourage them to blow paint

by dropping paint on the paper and blowing it using the straw. Compare the air coming through the straw with the wind.

- **BULLETIN BOARD**—Display pictures of how the wind helps people and how it can do harm. Discuss storms and tornadoes.

- **LARGE MOTOR**—Tie strips of crepe paper to the climber and wheel toys, or encourage children to run in the wind with crepe-paper streamers. Discuss what happens to the crepe paper when the wind blows through it.

- **MUSIC**—Play various kinds of music and provide scarves for children to dance in the wind.

- **OUTSIDE**—Fly a kite on the play yard. Compare different types of kites and kites made from different materials.

- **OUTSIDE**—Provide children with bubble solutions to blow bubbles in the wind.

- **SCIENCE**—Chart the various wind creations and how far they travel in the wind.

- **STORY**—Read *Gilberto and the Wind* by Marie Hall Ets (see appendix A on p. 177). In this story, Gilberto has a fun day seeing all the wind can do.

- **WRITING**—Encourage children to compare their various wind creations and how they respond to the wind. Be sure to provide science journals for children to record their observations.

# Shadow Hunt  Outside

## DID YOU KNOW?

People, plants, and animals make shadows when they come between the sun and the earth. Shadows change when the position of the sun changes or when the people, plants, or animals move. This activity will encourage children to experiment with making different shadows.

## MATERIALS

- **SUNNY DAY**
- **CHALK**
- **SIDEWALK**

## ACTIVITY

1. As children are playing outside during self-selected activity time, encourage them to notice their shadows. Ask them open-ended questions like these:
   - What makes a shadow?
   - How can you make your shadow change?
   - Where could you hide from your shadow?
   - What happens to your shadow when you go inside?
   - How is your shadow different now than it was in the winter?

2. Challenge children to trace their shadows on the sidewalk or a large piece of paper.

3. Encourage children to explore shadows on the play yard and discover their source.

Every year, children make a discovery either during the early spring or late fall that becomes the basis of this activity. We are outside late morning and the angle of the earth to the sun causes shadows to be very prominent. The exploration of shadows leads to discussions about the sun and how the angle changes with time of day and season. This leads to futher discussion about other things that change when shadows change—weather, animal behavior, or seasons. For this experience to be meaningful for children, teachers need to be aware of shadows at the time children are outside and have materials ready.

## ADDITIONAL ACTIVITIES

- **ART**—As children draw, challenge them to add shadows to people, plants, and animals in their artwork. Then discuss the time of day or year they are depicting.

- **BULLETIN BOARD**—Take pictures of children and their shadows or shadows of familiar objects on the play yard. Set up a display where children can match people or objects with their shadows.

- **GROUP**—Play "Whose Shadow?" by hanging a sheet in the classroom and placing a strong light behind it. Have the children hide their eyes while one child at a time goes behind the sheet. The rest of the children guess whose shadow they see.

- **LARGE MOTOR**—During self-selected activity time, let children play in the light of an overhead projector. Encourage them to experiment with different objects and the shadows they create.

- **MANIPULATIVE**—Play an animal and shadow matching game with animal pictures cut from magazines and their shadows or silhouettes cut from black paper. Put these out during self-selected activity time for children to match.

- **OUTSIDE**—Play "Lose Your Shadow." Encourage children to find shadows to hide in so they won't have a shadow, or show children how to play shadow tag.

- **SCIENCE**—During self-selected activity time, challenge children to find different things that might add color to their shadows (such as colored cellophane, colored plastic, or color viewers). Discuss how they think colored shadows are made.

- **SCIENCE**—Draw around shadows at different times of the day. Compare and record how the shadow changes over time.

- **STORY**—Read *The Shadow Book* by Beatrice Schenk DeRegniers (see appendix A on p. 187). This is a story of shadows and how they follow a child through the day.

# Where Do All the Dead Leaves Go? ✹ Group

## DID YOU KNOW?

When dead plants and animals decay, they become part of the soil and help improve it. This activity will help children explore the process of decay as well as experiment with materials that do not decompose.

## MATERIALS

- **LEAVES**
- **ORANGE PEEL**
- **ROCK**
- **STYROFOAM CUP**
- **SODA CAN**
- **SHOVEL**
- **CHART PAPER AND MARKER**
- **SCIENCE JOURNALS**

## ACTIVITY

1. On chart paper, create a chart with four columns and label each column: "Three Days," "One Week," "Two Weeks," and "One Month." Draw a picture of each object down the left side and list predictions under each column.

2. Select a spot on the play yard where children will be able to dig into the ground.

3. Encourage children to predict what will happen to the items if they are buried in the ground. Ask them open-ended questions like these:
   - What do you think will happen?
   - Why do some items change more than others?
   - What will happen to those materials that don't change?

   Record their predictions on the chart you prepared ahead of time. Provide science journals for children to document their thoughts and ideas.

4. Brainstorm other items the children might want to include in the experiment.

5. Have children bury the items, marking the locations so they can be dug up again in a few days.

6. After three days, have children dig up the items. Examine the items, then rebury them. Repeat the process again in a week, then two weeks, then a month.

## Ideas from Sherri's Classroom

This learning event came about as the result of a child's cat dying. The family had a ceremony for the beloved pet and buried it in the backyard. Each day, however, the preschooler asked if they could dig the cat up—the child was curious about what happened to the animal in the ground. The parent shared a concern about this behavior with me and I reassured her that this was typical of preschool thinking. However, to help this particular child learn about the process of decay, we started this experiment at school. Although I didn't think of it as a conservation experience when it began, it quickly turned into one when the children decided what things we might bury.

7. Each time the items are dug up, discuss the changes that have taken place and record the children's observations on the chart.

## ADDITIONAL ACTIVITIES

- **BULLETIN BOARD**—Take before, after, and in-between pictures of each of the objects. Display with the chart so children will have a visual reminder of what the objects looked like throughout the experiment.
- **GROUP**—Discuss the leaves on the ground in the spring and how they are decaying. Decay proceeds slowly when the temperature is cool but now that the warmth of spring is present, decay will proceed more quickly.
- **OUTSIDE**—Fence in a small area near your school and create a compost pile for your garden.
- **SCIENCE**—Encourage children to examine and explore leaf litter on the play yard. Talk about the changes in the leaves over time.
- **SCIENCE**—Collect a large bag of packing peanuts that are water soluble. Place in the sensory table with a large pitcher of water and big spoons. Have children predict what will happen when the water is stirred into the peanuts (avoid telling them before the experiment that the peanuts are water soluble). Repeat the experiment with nonsoluble peanuts. Chart their predictions and the results. This experiment can also be done with the peanuts placed in soil. After they dissolve, discuss which would be better for the earth and why.
- **STORY**—Read *The Dead Tree* by Alvin Tresselt (see appendix A on p. 188). The story demonstrates what happens to a tree when it dies.

# Explore a Log ✿ Science

## DID YOU KNOW?

Decomposed material is an important part of soil. This activity will enable children to explore how decomposed trees contribute to soil. It will also provide an opportunity to explore animal habitat in a rotting log.

## MATERIALS

- ROTTING LOG OR TREE STUMP
- SMALL CHILD'S WADING POOL OR LARGE TUB
- STICKS OR PROBES
- TWEEZERS OR FORCEPS
- MAGNIFYING GLASSES
- SCIENCE JOURNALS
- COLORED PENCILS
- CAMERA (OPTIONAL)
- TAPE RECORDER (OPTIONAL)

## ACTIVITY

1. Place the rotting log or tree stump in the tub or wading pool in the science area. Be sure the probes, tweezers, magnifying glasses, science journals, and colored pencils are available for children to use as needed. Ask: "What do you think might live in a rotting log?"

2. During self-selected activity time, invite children to explore the log. Record their discoveries, using photographs, audio recordings, drawings, and charts as documentation.

3. Encourage children to handle the decayed material. Have them squeeze the log debris and talk about the water in it. Look for animals that live in the log and talk about their job in soil-making. Include hand washing as part of the project. As you watch children explore the log, ask them open-ended questions like these:
   - How did these animals get here?
   - How is this log helping the soil?
   - How do the animals help the soil?
   - What will happen to the log?

One early spring we visited a lake near our school where we found several rotting logs on our travel through the woods. Although we explored the logs in their natural habitat, the children decided they needed to take one back to the classroom for futher investigation. The experience led to much discussion and speculation as to what happens over time to trees. The ecosystem created by the tree when it was living was compared to the ecosystem created by the tree after it died.

After thoroughly examining the log, placing it in an area of the play yard where it could be futher explored allowed the children to consider what happened to the log over time.

4. Display children's documentation to encourage further avenues of study and to share their experience with parents.

## ADDITIONAL ACTIVITIES

- **FIELD TRIP**—Explore a rotting log in the woods. Compare the findings on the field trip with those found in the classroom. Be sure to photograph the two experiences for documentation and comparison.
- **OUTSIDE**—When children begin to lose interest in the activity, place the rotting log in an isolated area of the play yard. Revisit the log periodically throughout the year. Discuss changes in the log over time.
- **STORY**—Read *Once There Was a Tree* by Natalia Romanova (see appendix A on p. 188). This story explores the life cycle and ownership of a tree.

# Mud Pies ✿✿ Outside

## DID YOU KNOW?

Soil consists of three types of rock particles—clay, silt, and sand. Clay particles are fine and light in color. When water is added, clay becomes sticky. Silt particles have the consistency of flour. Sand particles feel gritty. Loam soil is a mixture of sand, silt, and clay. Humus soil is dark and loose and consists of decayed remains of plants and animals. This activity will help children understand that there are different types of soil.

## MATERIALS

- **SHOVELS**
- **PIE PANS OR PAPER PLATES**
- **INDIVIDUAL MAGNIFYING GLASSES (ONE FOR EACH CHILD)**
- **WATER**
- **CHART PAPER AND MARKER**

## ACTIVITY

1. During group time, discuss the children's various recipes for mud pies.

2. Record their ideas on chart paper.

3. During self-selected activity time outside, set out the shovels, pie pans or paper plates, magnifying glasses, and water near an area where children can dig. Ask them open-ended questions like these:
   - How does the soil feel before you add water? How does the soil feel after?
   - What does your soil look like under the magnifying glass?
   - What can you find to add to your mud pie?
   - What makes mud pies look different from one another?
   - Why did this mud pie harden faster than that one?

4. Encourage children to experiment with making mud pies and decorating them with various items found in the environment.

## ADDITIONAL ACTIVITIES

- **ART**—During self-selected activity time, encourage children to make different colors of paint by mixing water with different-colored soils. Invite children to try different natural objects to paint with (such as feathers, sticks, or grass) and on (such as rocks, leaves, or bark).
- **FIELD TRIP**—Locate an area where clay is easily accessible. Encourage children to dig clay and take it back to the classroom. In a manual food processor, process the clay to grind small sticks and rocks and to uniformly disperse moisture. Knead and shape into desired forms. Allow to dry, then take to a local ceramics shop or build a kiln and fire on the parking lot (Kohl and Gainer 1991, 204).
- **SCIENCE**—Ask children to bring soil from their backyards. Encourage them to compare these types of soil with the samples taken from the play yard. Provide science journals for them to record their ideas.
- **SCIENCE**—Provide different types of soil for children to make mud pies. Notice textures, colors, and smells. Place some mud pies in shady spots and some in the sun. Observe which mud pies dry the fastest.
- **STORY**—Read *Muddigush* by Kimberly Knutson (see appendix A on p. 187). The story provides a delightful texture of sounds for playing with and in the mud.
- **WRITING**—Provide materials for each child to contribute a journal page or recipe to a class mud-pie book. Create the children's own "book on tape" by making an audio recording of their work.

# What Lives in the Soil?  Outside

## DID YOU KNOW?

Soil is the home for many small animals. These animals help the soil by eating bits of decayed plants and animals. This activity will enable chil-dren to discover some of the animals that live in soil.

## MATERIALS

- **SOFT EARTH**
- **HAND SHOVELS**
- **TWEEZERS OR TONGS**
- **MAGNIFYING GLASSES (ONE FOR EACH CHILD)**
- **SCIENCE JOURNALS**

## ACTIVITY

1. During self-selected time outdoors, encourage children to carefully explore the soil for animals that might be living there. Ask them open-ended questions like these:
   - How can you tell if an animal has been here?
   - How do you think these animals help the forest?
   - What do you think these animals eat?

2. Study the soil with magnifying glasses. Encourage children to document their observations in their science journals.

## Additional Activities

- **Art**—Provide socks and various accessories for children to make soil creatures during self-selected activity time.
- **Pretend play**—During self-selected time, provide tubs of soil with plastic worms, insects, snails, centipedes, and other animals that might live in the soil for children to play in and explore.
- **Pretend play**—Using a projector, show an overhead of soil to the class. Provide children with antennae, socks, wings, noses, and so on to create shadows of the various animals moving about in the soil.
- **Science**—Create a worm ranch with the children. Cover the bottom of a large, clear jar with rocks. Layer soil and leaves in the jar until three-quarters full. Add worms (Red River Wigglers work well) and a teaspoon of coffee grounds for worm food. Keep soil moist but not wet and cover the outside of the jar with dark paper. Check for worm tunnels every few days. Talk about how the worms help soil by decomposing debris, mixing, aerating, and fertilizing it.
- **Science**—Prepare a spot to see earthworm tracks. Pour a bucket of water over some soil in the play yard. Make the soil muddy, smooth it, and then leave it alone. The next day, search for worm tracks. Challenge children to make different habitats for worms. One might include sand, another might include clay. One might be wet, another dry. Keep daily records of the worms' responses to the habitats.
- **Story**—Read *Deep Down Underground* by Oliver Dunrea (see appendix A on p. 187). This counting story illustrates earthworms, beetles, ants, spiders, and other creatures marching, burrowing, scurrying, and "scooching" deep down underground.

# How Does Your Garden Grow? ✿ Science

## DID YOU KNOW?

Spring rains and melting snow add moisture to the soil and cause seeds to swell. The sun's warmth helps the seeds to grow into small plants. This activity will provide insight for what plants need in order to grow.

## MATERIALS

- EMPTY EGG CARTON WITH THE TOP CUT OFF (ONE PER CHILD)
- FLOWER SEEDS (PROVIDE SEVERAL VARIETIES)
- POTTING SOIL
- WATER AND WATERING CONTAINERS
- SCIENCE JOURNALS

## ACTIVITY

1. Place the materials in the science center during self-selected activity time.

2. Encourage children to place potting soil in several sections of an egg carton, leaving at least two of the sections empty.

3. Allow children to add seeds to each of the sections, including those without soil.

4. Provide water for children to water their seeds.

5. Place the egg cartons around the classroom. Be sure to put several in dark areas and some in strong sunlight.

6. Discuss children's ideas about what will happen with the seeds. Ask them open-ended questions like these:
   - How long do you think we'll have to wait before anything happens?
   - Why did some plants grow better than others?
   - What do plants need to grow?

7. Check the seeds daily and record the children's observations. Place science journals near plantings so children can record their daily observations.

## Additional Activities

- **Bulletin board**—Cover the bulletin board with brown paper. Place strips of Velcro in rows on the paper. Provide pictures of indigenous garden fruits, vegetables, and flowers with Velcro attached for children to arrange in the garden.

- **Field trip**—Take a field trip to a wooded area to look for different kinds of mosses growing on or near rocks and tree trunks. Discuss the various textures and colors, and talk about where the mosses are growing and their habitat. If collected in an unprotected area, samples might be taken to grow a moss garden in the classroom.

- **Outside**—Allow children to move toy tractors and farm equipment to an outside dirt or sand pile during self-selected activity time.

- **Pretend play**—Put out gardening gloves, hats, tools, and empty seed packets in the dress-up area for children to use during self-selected activity time.

- **Science**—Encourage children to bring soil from home. Place the different soils in each section of an egg carton and record the results of planting seeds in different soils.

- **Science**—Provide children with materials to try other types of planting. Choose from the following:
  - Substitute empty eggshells for the egg cartons.
  - Provide bean seeds, damp paper towels, and resealable plastic bags. Encourage children to place all the materials in the bag, seal it, and tape it to the window.
  - Have each child bring an old tennis shoe from home and plant a tennis-shoe garden.
  - Start a sweet-potato vine.
  - Plant a garden in the play yard.

- **Science**—Provide different liquids (such as vinegar, lemon juice, and so on) for children to "water" the plants. Keep a daily record of the observable effects of each liquid on the plants.

- **Story**—Read *Planting a Rainbow* by Lois Ehlert (see appendix A on p. 187). This is a simple story of a parent and child planting bulbs, seeds, and young plants to grow a rainbow of flowers.

# What Is Erosion? 🐁 Group

## DID YOU KNOW?

Erosion is caused by wind and rain on soil with no plants to hold it in place. Erosion causes precious topsoil, which is essential for growing things, to end up at the bottom of hills, streams, rivers, and oceans. This activity will demonstrate soil erosion.

## MATERIALS

- **BLOW DRYER**
- **HOSE OR BUCKET OF WATER**
- **PILE OF DRY SAND OR SOIL**
- **SCIENCE JOURNALS**

## ACTIVITY

1. Gather children outside around the soil or sand, and have them form a hill.

2. Pretend that the hill has just been plowed by a farmer. A big storm is coming and a strong wind is blowing. Ask children open-ended questions like these:
   - What do you think will happen to the soil when the wind blows? When the heavy rains fall?
   - Can you find any places on our play yard where the soil is eroded? How could we fix it?
   - Why is erosion a problem for people and animals?

   Encourage children to document in their science journals what they predict might happen.

3. Simulate wind with the blow dryer. Note what happens. Try the blow dryer on soil where the grass is growing. Observe the difference.

4. Repeat the experiment using water from a hose or bucket.

## ADDITIONAL ACTIVITIES

- **BULLETIN BOARD**—Photograph the experiment and display the photos along with children's thoughts and ideas about soil erosion. Include pictures of actual soil erosion.
- **FIELD TRIP**—Look for erosion on a field trip while walking around the neighborhood.

# Flower Shop ✿ Pretend Play

## DID YOU KNOW?

Flowers are a beautiful part of the spring landscape throughout the United States. There are many different types of flowers—both wild and garden varieties. This activity will expose children to a variety of flowers as well as how people use this natural resource.

## MATERIALS

- **ARTIFICIAL FLOWERS**
- **ASSORTED FLOWER POTS AND VASES**
- **PICTURES OR BOOKS OF FLOWERS AND FLOWER ARRANGEMENTS**
- **CASH REGISTER**
- **TELEPHONE**
- **NOTE PAD**
- **PENCILS**

## ACTIVITY

1. Set up a flower shop in your pretend-play area during self-selected activity time.

2. Encourage children to experiment with flower arranging as they play flower shop. Ask them open-ended questions like these:
   - Where do flower shops get their flowers?
   - Which flower do you like best?
   - Where do you think we could find a flower like this?
   - Why do we need flowers?

## ADDITIONAL ACTIVITIES

- **ART**—Place bulb or seed catalogs, glue or glue sticks, construction paper, and scissors in the art center during self-selected activity time. Encourage children to create flower collages, cards, or other creations.

- **BLOCK**—Place silk flowers and plastic vegetables in the block area for children to pretend to plant and harvest in a garden during self-selected activity time.

- **FIELD TRIP**—Visit a nature area to observe spring wildflowers in their native habitats. Be sure to take along clipboards, paper, and colored pencils so children can sketch the flowers rather than pick them. Discuss the differences and similarities between wildflowers and those grown in the garden.

- **MANIPULATIVE**—Create a matching game using pictures (or stickers) of flowers native to your area. During self-selected activity time, place the flower matches in the manipulative area. Encourage children to place the cards face down and create memory matches or play a card game such as "Old Maid."

- **PRETEND PLAY**—Place soil, flowers (artificial or real), gardening tools, and gardening gloves in the sensory table for children to experiment with during self-selected activity time.

- **STORY**—Read *Counting Wildflowers* by Bruce McMillan (see appendix A on p. 185). This counting book features photographs of common wildflowers.

# Dandelion Potpourri ❀ Group

## DID YOU KNOW?

Dandelions grow wild in yards, fields, and play yards across the country. Children delight in the appearance of these yellow flowers in the spring and often pick bou- quets of them to share with special people. This experience will encourage children and grown-ups to appreciate the dandelion in a different way.

## MATERIALS

- **DANDELIONS**
- **FRAGRANT SPICES (SUCH AS CINNA- MON STICKS, CLOVES, OR ROSEMARY)**
- **DRIED ORANGE OR LEMON PEEL**
- **DRIED REINDEER MOSS (THIS IS A FIX- ATIVE FOR THE FRAGRANCE AND CAN BE OMITTED IF UNAVAILABLE; HOW- EVER, FRAGRANCE WILL NOT LAST AS LONG)**
- **ESSENTIAL OIL (FOUND AT MOST CRAFT STORES)**
- **FABRIC SQUARES**
- **RIBBONS**
- **EMPTY JET-DRY DISHWASHER BASKETS**
- **GLASS OR METAL BOWL**

## ACTIVITY

1.  As children pick bouquets of dandelions in spring, encourage them to remove flowers from the stems and spread them out on trays to dry. Depending on the humidity, allow two to three weeks to dry.

2.  Once the dandelions have dried, place them in a glass or metal bowl. Allow children to add spices, orange or lemon peel, and any other dried flow- ers or leaves they have collected.

3.  Have children measure approximately one tea- spoon dried reindeer moss for every two cups of potpourri. (This is a fixative and will hold the fragrance of the potpourri.) Add this to the flower and spice mixture, and stir.

4.  Next, have children add several drops of the essential oil into the potpourri.

5.  Provide squares of fabric in which to place pot- pourri. These can be tied with ribbon or placed in Jet-Dry dishwasher baskets. Ask children open-ended questions like these:
    - What else can you do with dandelions?
    - What are dandelions good for?
    - Why do some people dislike dandelions?

## Additional Activities

- **Art**—During self-selected activity time, show children how to use dandelions and other flowers and leaves to create pictures by rubbing, pressing, and squeezing them on white paper.
- **Large motor**—Encourage children to pretend to be a class dandelion. As a group, pretend to grow, flower, turn to seed, and then scatter with the wind. Finish by growing once again.
- **Outside**—Encourage children to blow on dandelion seed heads and follow the seeds to see where and how far they go.
- **Story**—Read *The Dandelion Seed* by Joseph Anthony and Cris Arbo (see appendix A on p. 187). This beautifully illustrated book shows the travels and life cycle of a dandelion seed.
- **Woodworking**—Provide materials for children to create a simple flower press. Drill a hole through each corner of two wood squares, each ¼-inch thick. Children will need a long bolt and wingnut for each of these corners. Use cardboard to layer flowers between the two boards and bolt together. Leave for several days or until flowers are dry.

# Seeds, Roots, and Plants ❧ Story

### DID YOU KNOW?

Seeds provide food for animals and develop roots that hold the soil in place. This story will help children realize all seeds are not alike. Different seeds grow different plants, and seeds are transported by the wind or animals. This activity will demonstrate how the roots and stem of the plant grow from the seed as well as ways seeds travel.

## MATERIALS

- **SPRING PATTERN 3 (SEE APPENDIX D, PP. 232–233)**
- **FELT CHARACTERS**
- **FLANNELBOARD**

## ACTIVITY

1. Read the following story to children, placing felt characters on the flannelboard at the appropriate times.

2. After the story reading, ask children open-ended questions like these:
   - How do you think seeds know when it's time to grow?
   - What do seeds need to grow?
   - How do seeds know what kind of plant to become?
   - How does the soil move if there isn't any ice or snow?

## ADDITIONAL ACTIVITIES

- **MANIPULATIVE**—Make a matching game of seeds and their plants by placing several different kinds of seeds on posterboard and covering with clear Con-Tact paper. Write the name of the seed under each type. Use the seed packets (also covered with Con-Tact paper) for children to match with the seeds. Place in the manipulative area during self-selected activity time.

- **MANIPULATIVE**—Create a plant puzzle that illustrates various plant parts (roots, stem,

Once there was a clover plant.
On the plant were leaves.
With the leaves were little stems
and on the stems were seeds.

Lots of seeds of similar size
crowded on each stem.
The clover plant had made the seeds
and one seed's name was Jim.

Along came a hungry deer
eating clover leaves.
He almost gobbled Jim up,
as he munched on clover seeds.

The green leaves on the clover plant
began to turn to brown.
As winter came the clover seeds,
fell off onto the ground.

Along came a hungry quail
pecking at the ground.

Eating up the clover seeds,
every one she found.

Peck, peck, she ate the seeds.
Look! Peck! Search! Peck! Seek!
The quail saw Jim and pecked him up
with her shiny beak.

Up, away flew the quail
with Jim in her beak.
Higher than the trees she flew
across a field and creek.

When she landed in a field
she dropped Jim on his back.
Bump! Bump! Bounce! Bounce!
Jim rolled into a crack.

Rain and ice and melting snow
moved the dirt around.
By the time that spring arrived,
Jim was underground.

Sun shone from the sky above,
then rain and morning dew.
Jim began to swell and grow.
Jim split right in two.

Up out of the moist damp dirt,
a sprout came out of Jim.
Down more deeper in the soil,
a root grew out of him.

Up out of the soil he grew
and leaves came from his stems.
Soon he was a clover plant
and flowers grew on him.

The rain and sun and fertile soil
give Jim the things he needs.
The flowers growing on his stems
slowly turned to seeds.

—John Griffin

flower, and leaves) out of felt for use at the flannelboard.

- **SCIENCE**—Place white carnations or Queen Anne's lace and water with food coloring in the science area during self-selected activity time. As children become more curious, ask them what they think will happen when the flowers are placed in the colored water. Record their ideas and have children place flowers in the water to observe throughout the day.
- **SCIENCE**—Force buds to bloom early by placing forsythia or pussy-willow branches in water.
- **STORY**—Read *The Reason for a Flower* by Ruth Heller (see appendix A on p. 185). This nonfiction story illustrates the various plant parts and the role of flowers.

# Let's Go Camping! ⚅ Pretend Play

### DID YOU KNOW?

Throughout the country camping is an activity enjoyed by people of all ages. Although the more rugged among us continue to camp throughout the colder months of the year, most families with young children prefer to camp from early spring until late fall. This activity will allow children to see another way people enjoy our nation's vast natural resources.

## MATERIALS

- **TENT**
- **SLEEPING BAGS**
- **CAMPING DISHES**
- **LOGS OR STUMPS (OPTIONAL)**
- **BINOCULARS**
- **COMPASSES**
- **OTHER CAMPING EQUIPMENT THE CHILDREN MIGHT THINK OF**

## ACTIVITY

1. Set up a campsite either inside the classroom or outside in the play yard (or both). Allow children to play with and explore the equipment during self-selected activity time. Ask them open-ended questions like these:
   - What do people need to go camping?
   - How does camping affect the wildlife in the area?
   - What kinds of animals and plants do you see when you go camping?
   - How do you know where it is safe to camp?
   - How do you care for a campsite?
   - How do you know what the camping rules are in a particular area?

2. As children become more interested in the campsite, talk about other items that might be added (for example, stuffed animals that represent animals indigenous to your region of the country, stones to place around a pretend campfire, branches to represent trees, or audio recordings of bird, frog, insect, or nature sounds).

This is always a favorite activity for the children. It is generally initiated by someone going camping with their family and often remains a part of our classroom for a month or more. Even children who have no camping experience develop a healthy attitude and warm feelings about this common use of our natural resources. Some families have even decided to take their children camping as a result of this activity.

## ADDITIONAL ACTIVITIES

- **ART**—Large animals can be dangerous, especially when startled. One method campers often use to help animals know people are in the area is a noisemaker. These can be made from recycled cans filled with gravel and closed with plastic lids or duct tape. Children can decorate the cans to make them more attractive. This is an excellent opportunity to talk with children about disturbing animals in their homes.

- **FIELD TRIP**—Visit both a natural camping area and a more commercial area. Compare the experiences people would have camping in the two areas. Look for posted rules and regulations for area use. Talk with children about following the rules.

- **GROUP**—Invite an outdoor specialist to demonstrate starting a campfire with flint.

- **GROUP**—Invite someone who owns a recreational vehicle to visit the class. Discuss the difference between RV and tent camping.

- **GROUP**—Brainstorm a list of materials that might be needed in a first-aid kit. Help children gather materials and prepare the kit.

- **MANIPULATIVE**—Provide long shoelaces or rope for children to experiment with tying knots. Discuss how important knots are for camping.

- **NUTRITION**—Prepare trail mix from cereal, nuts, pretzels, raisins, and seeds to take on a camping trip. Discuss how campers keep their food fresh without coolers.

- **OUTSIDE**—Safely start a campfire and cook s'mores or hot dogs. Demonstrate how to safely extinguish the fire.

- **OUTSIDE**—Introduce the use of compasses to children. Locate north and talk about that constant with the compass. Often giving children a simple association, such as polar bears at the North Pole or penguins at the South Pole, helps them make a concrete connection. Challenge them to find north in several different places both in and outside the classroom.

- **WRITING**—Provide pencils, papers, compasses, and so on for children to create maps of the campground or a hiking trail.

# What's an Insect? ✾ Science

## DID YOU KNOW?

Over half of all living things in the world are insects. Insects have six legs, three body parts, a pair of antennae, and are covered with chitin, a light, flexible, waterproof substance.

Most insects have wings. Insects are an important part of many food chains. This activity will allow children to identify some of the characteristics of insects.

## MATERIALS

- LIVING, DEAD, OR PLASTIC INSECTS, SUCH AS A GRASSHOPPER, CRICKET, COCKROACH, CATERPILLAR, BUTTERFLY, LADYBUG, OR DRAGONFLY (SELECT COMMON INSECTS, INDIGENOUS TO YOUR AREA IF POSSIBLE)
- MAGNIFYING GLASSES
- TWEEZERS
- SCIENCE JOURNALS
- COLORED PENCILS

## ACTIVITY

1. Place the insects and the magnifying glasses in the science area during self-selected activity time.

2. Encourage children to examine the insects and discuss children's observations. Ask them open-ended questions like these:
   - What do all of these insects have that make them alike?
   - What do insects eat?
   - Where do insects live?
   - How do insects help people? How do they hurt them?
   - Who are the insects' enemies?

   Provide science journals and colored pencils for children to document their observations.

## Additional Activities

- **Art**—Provide Styrofoam pieces, cut-up egg cartons, construction paper, pipe cleaners, toothpicks, and other materials for children to create insects during self-selected activity time.
- **Bulletin board**—Create a display of insect pictures along with children's comments about insect characteristics.
- **Field trip**—Take children on a hike to look for insects. Make note of the insects' habitat.
- **Large motor**—Encourage children to study how various insects move, and then try moving like them.
- **Large motor**—During group time, discuss how bees communicate with other hive members through dance. Encourage children to make up dances that will communicate with other children where snack is located, or how to get to a secret treasure on the play yard.
- **Outside**—Encourage children to be quiet and listen for insect sounds.
- **Science**—Have the children observe grasshoppers or crickets jumping. Challenge children to estimate how far and how high they think the grasshoppers or crickets will jump.
- **Story**—Share the book *What's Inside? Insects* by Angela Royston (see appendix A on p. 183). This book explores the parts of an insect and the functions of those parts.

# My Friend, Little Caterpillar &#x2042; Music

## DID YOU KNOW?

Butterflies and moths go through several stages of development. However, moths are different than butterflies. They have thicker bodies and are usually larger. When moths rest their wings, they spread them out; butterflies keep their wings folded above them while they rest. Butterflies have little knobs on the antennae. This song will help children relate to the changes butterflies and moths go through, as well as similarities and differences between butterflies and moths.

## MATERIALS

- **PICTURES OF BUTTERFLIES AND MOTHS**
- **LIVING, DEAD, OR PLASTIC BUTTERFLY AND MOTH**
- **SCIENCE JOURNALS**

## ACTIVITY

1. Teach children the following song to the tune of "Twinkle, Twinkle Little Star":

   My friend little caterpillar (*Pet thumb.*)
   Eats green leaves until they fill her.
   Then she grows a chrysalis. (*Make fist with thumb in middle.*)
   There she sleeps unseen by us.
   Spring will come and by and by
   she'll become a butterfly. (*Put thumbs together and flap hands.*)

   My friend little caterpillar (*Pet thumb.*)
   Eats green leaves until they fill her.
   Then she spins a small cocoon. (*Make fist with thumb in middle.*)
   There she sleeps and very soon
   Spring will come and she'll fly off. (*Put thumbs together and flap hands.*)
   She'll become a fuzzy moth.

2. Place moth and butterfly, as well as the pictures, in the science area during self-selected activity time. Encourage children to examine both the moth and the butterfly, and record their observations in their science journals. Ask them open-ended questions like these:

   - How do the moth and butterfly look the same? Different?

- How do they act alike? Different?
- How much does a caterpillar need to eat before it forms a chrysalis or cocoon?
- How does the butterfly or moth know when to come out of the chrysalis or cocoon?
- How does the caterpillar know whether to grow a chrysalis or spin a cocoon?
- Why do you think butterflies and moths are so many colors?

## ADDITIONAL ACTIVITIES

- **ART**—During self-selected activity time, choose from one of the following:
  - Provide materials for children to create folded paper paintings by making a crease down the center of a sheet of paper, painting on one side of the crease, and folding and opening the paper so a symmetrical design appears. Compare their work to the symmetry of a butterfly or moth's wings.
  - Provide clipboards, colored pencils, watercolors, crayons, and markers for children to sketch butterflies and moths they see or capture on the playground.
  - Provide egg cartons, cardboard tubes from toilet paper rolls, pipe cleaners, tissue paper, clothespins, and other materials for children to create caterpillars, butterflies, and moths. Be sure to let them create the designs for themselves rather than providing a model.
- **LARGE MOTOR**—Pantomime the stages of metamorphosis of the butterfly or moth.
- **OUTSIDE**—Look for butterflies and moths on the play yard during self-selected time.
- **OUTSIDE**—Create butterfly nets from wire hangers. Place these outside for children to try to catch butterflies and moths during self-selected activity time. Study and sketch the specimens caught by the children, but be sure to release them at the end of the day.

- **PRETEND PLAY**—During self-selected activity time, provide party blowers for children to pretend to eat like a butterfly or moth. Add headband antennae and wings made from coat hangers and pantyhose.
- **SCIENCE**—Bring a caterpillar to class for children to observe. Encourage children to write about their observations and sketch the caterpillar in their science journals. Be sure to have the children set the caterpillar free at the end of the day.
- **STORY**—Locate a copy of the book *Eyewitness Juniors: Amazing Butterflies and Moths* by John Still (see appendix A on p. 183) and display it in your reading center. This nonfiction book uses photos to illustrate the life cycles and characteristics of various kinds of moths, butterflies, and caterpillars.
- **WRITING**—Provide pencils, stiff paper, and crayons for children to write butterfly or moth stories during self-selected activity time. This should be a free choice experience, allowing children to spend as much time or as little time with it as they feel is necessary. Document children's conversations, and bind the pictures, stories, and documentations together to create a class book about butterflies and moths.

# Who Am I? 🎵 Music

## DID YOU KNOW?

The rabbit is a very common wild animal. It eats green plants and is prey for many animals. It usually lives in brush rather than holes in the ground. Rabbit tracks show the long footprints of the rabbit's hind feet in front of the small prints of the front feet.

The rabbit touches the ground with his front feet first, then swings its hind feet forward past the front feet before touching the ground. This activity will help children learn about rabbits and experience how they move.

## MATERIALS

- PICTURE OF RABBIT TRACKS
- RABBIT PICTURES
- AUDIO PLAYER
- LIVELY MUSIC

## ACTIVITY

1. Teach children the following song to the tune of "Three Blind Mice":

   Long, skinny ears. (*Hold up fingers behind head for ears.*)
   A little white tail. (*Turn around and wiggle tail.*)
   I live on the ground (*Point to ground.*)
   and I hop all around. (*Hop two fingers around.*)
   I have big eyes so I can see, (*Put fingers around eyes.*)
   coyotes and foxes who'd like to eat me. (*Look around with hand over eyes.*)
   I nibble on green plants that I see. (*Pretend to nibble.*)
   Who am I?

2. Show children the picture of the rabbit tracks. Point out the unusual nature of the tracks and discuss how rabbits get their hind feet around and ahead of their front feet. Ask children open-ended questions like these:

- Where would be a good place for a rabbit to live?
- What do you think a rabbit would like to eat?
- Who are the rabbit's enemies?
- Why is it so hard for people to get their feet in front of their hands, but it looks so easy for rabbits? How do rabbits learn to hop like this?

3. Demonstrate how a rabbit hops. Play lively music and encourage children to practice doing the bunny hop.

## ADDITIONAL ACTIVITIES

- **LARGE MOTOR**—Have "bunny hop" races.
- **NUTRITION**—Serve "rabbit food" for snack: lettuce, spinach, celery, radishes, cauliflower, and broccoli. Remember, rabbits prefer young and tender green plants (carrot tops rather than carrots).
- **OUTSIDE**—Go on a "bunny hop" walk in the play yard.

# What Hatches from an Egg? ❊❧ Art

## DID YOU KNOW?

Birds, amphibians, reptiles, fish, insects, and spiders hatch from eggs, but only a few mammals actually hatch from eggs. This activity will help children learn about animals that hatch from eggs.

## MATERIALS

- **PICTURES OF ANIMALS THAT HATCH FROM EGGS**
- **LARGE PLASTIC EGG (ONE PER CHILD)**
- **STYROFOAM BALLS (VARIOUS SIZES)**
- **STYROFOAM PACKING**
- **PIPE CLEANERS**
- **FEATHERS**
- **POM-POMS**
- **SCRAPS OF CONSTRUCTION PAPER**
- **SEQUINS**
- **GLUE**
- **TOOTHPICKS**

## ACTIVITY

1. Place materials in the art area during self-selected activity time.

2. Discuss animals that hatch from eggs. Ask children open-ended questions like these:
   - How do you know which animals hatch from eggs?
   - How does the animal inside the egg know when to come out?
   - How do animals get out of the shell?

3. Encourage children to create some type of animal that hatches from an egg.

4. As children finish, suggest they place their animal in their egg so it can hatch.

## ADDITIONAL ACTIVITIES

- **ART**—Encourage children to create nests using twigs, branches, and mud during self-selected activity time.
- **FIELD TRIP**—Visit a hatchery.

- **GROUP**—Play "Mother Bird." One of the children is the mother bird, the rest are the babies. The babies hide throughout the designated area and peep softly until the mother bird gathers them all back into the nest.
- **GROUP**—Discuss with the group how parents in the wild care for their young. A baby bird out of the nest will still be cared for by its parents and doesn't need human assistance. Young wild animals should always be left in the wild as their parents are probably nearby.
- **MANIPULATIVE**—During self-selected activity time, choose from one of the following:
  - Play a matching game with bird nests by reproducing and laminating various nests and the birds who make them. Encourage the children to match the birds to their nests.
  - Encourage children to sort pictures of animals into those that hatch from eggs and those that don't.
  - Play a memory game with eggs by drawing different kinds of eggs in pairs on cards.
- **NUTRITION**—Make "bird nests" for snack with ⅓ cup honey, ½ cup brown sugar, ¾ cup peanut butter, 1 teaspoon vanilla, 3 cups chow mein noodles, and 1 cup coconut. Mix and heat until easily stirred. Use approximately one heaping tablespoon of mixture to form a nest. Use jelly beans, grapes, or puffed rice for eggs. Talk about the different kinds of nests that the different species of birds create.
- **NUTRITION**—Serve gummy worms for snack. Encourage children to be baby birds and the mother bird who returns to the nest to feed her babies.

- **OUTSIDE**—Help birds with their nest building by draping brightly colored yarn, string, and ribbon over a pinecone and hanging it in a tree. Later, look for these bright colors in bird nests found in nearby trees.
- **OUTSIDE**—Encourage children to look for bird nests on the play yard. Be careful not to disturb the nest, eggs, or adults, but observe the progress from a distance.
- **SCIENCE**—Bring in different types of eggshells and display them in the science area.
- **STORY**—Read *Chickens Aren't the Only Ones* by Ruth Heller (see appendix A on p. 179). This nonfiction book illustrates the many animals that hatch from eggs.

# Do You Smell My Mother? Group

## DID YOU KNOW?

Most mammals have a keen sense of smell. This sense of smell helps them to find mates, identify food, locate places to live, know when enemies are approaching, and identify their young. Unlike most mammals, people have a relatively poor sense of smell. This experience will enable children to relate to one way mammals use their sense of smell.

## MATERIALS

- EMPTY FILM CANISTERS (ONE PER CHILD)
- SELECTION OF FRAGRANCES (CHOOSE SCENTS THAT ARE PLEASANT TO YOUNG NOSES, AND BRING ENOUGH FOR HALF OF THE CLASS)

## ACTIVITY

1. Place a different scent in each of two film canisters, making pairs of smells.

2. Separate the pairs of smells—one group for the "babies" and one for the "mothers."

3. During group time, divide children into two groups. Explain that half of the children will be the mothers and half will be the babies.

4. Distribute one set of canisters to the mothers and one to the babies. Challenge the babies to find their mothers by matching their fragrances. Reverse the process by having the mothers find their babies.

5. Ask children open-ended questions like these:
   • How can you keep track of your smell without getting it mixed up with the others?
   • How does a mother mammal remember her baby's scent?
   • What smells do you like to remember?

## ADDITIONAL ACTIVITIES

• **BLOCK**—Place stuffed or plastic mother and baby mammals in the block area. Encourage children to build habitats for them during self-selected activity time.
• **FIELD TRIP**—Visit a farm where baby animals (such as calves, chicks, kittens, puppies, rabbits, pigs, horses, and so on) are being born or raised. Talk with the farmer about how the mother and young recognize each other.
• **MANIPULATIVE**—Make a matching game where children match pictures of mother animals with their young. Make this available during self-selected activity time.
• **NUTRITION**—Encourage children to close their eyes while snack or lunch is served. Ask children if they can identify the snack or lunch items by their fragrances.
• **SCIENCE**—Make a smell matching game by having the children match a different smell with its corresponding picture. Make this available during self-selected activity time.

• **STORY**—Read *Animals Born Alive and Well* by Ruth Heller (see appendix A on p. 184). This brightly illustrated nonfiction book introduces children to mammals.

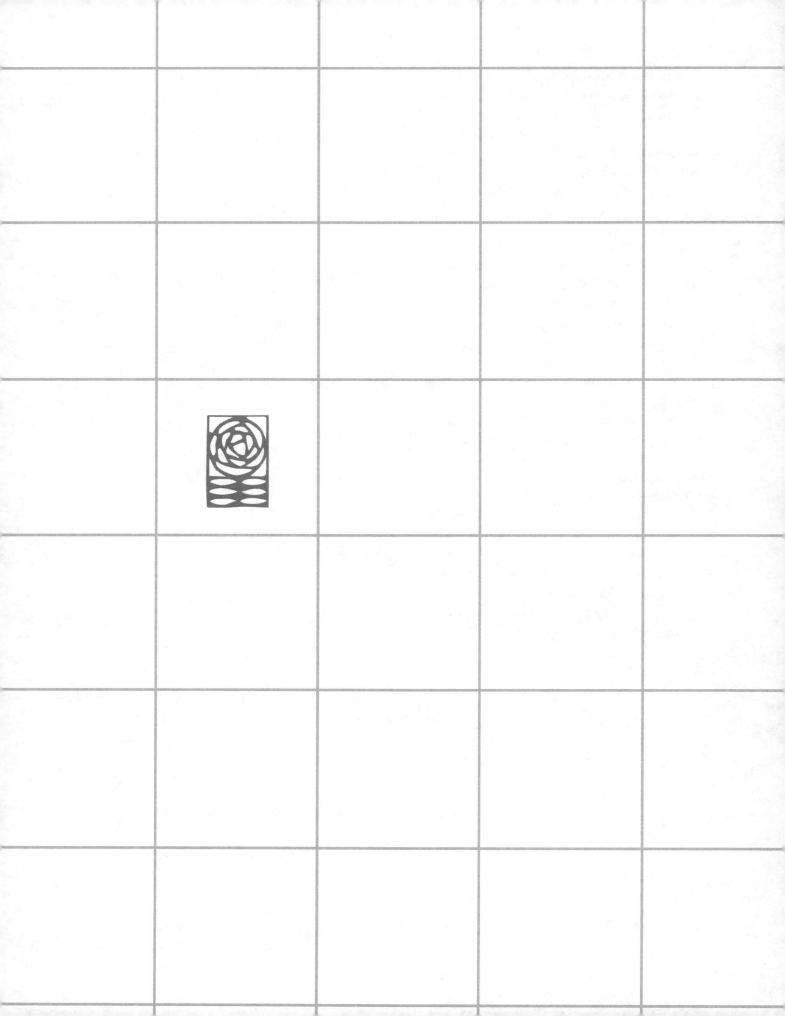

# Summer

The warm, sunny days of summer hold many opportunities for celebration. As the rotation of the earth provides more daylight, families enjoy outdoor activities, such as vacations, picnics, gardening, fishing trips, swimming, canoeing, and other recreational activities. Summer is the time to take advantage of the rich recreational areas across the country.

Young wildlife grow and develop during the summer. Gamebirds and songbirds help keep the insect population under control. Insects live and feed on the abundant plant life. By the end of summer, most young mammals and birds will exist in a self-sufficient lifestyle similar to their parents.

The Southeast experiences its rainy season in the summer, which is also the heart of hurricane season. In Everglades National Park, the summer's rainy season often brings wildfires sparked by lightning. Summer is the dry season in the West. As summer heat and wind dries out the forests, prime wildfire season begins.

Farmers usually harvest their wheat crops mid to late summer. Corn crops have been cultivated and are growing. Haymaking is in full swing.

The activities in the summer section of *My Big World of Wonder* take full advantage of the many outdoor opportunities. Topics include the following:

- Spiders
- Insects
- Pollution
- Fish
- Aquatic life
- Food chains
- Predator and prey relationships
- Rocks and minerals

# What Has Eight Legs? ▣ Science

## DID YOU KNOW?

A spider is not an insect but an arachnid. So are mites, ticks, and scorpions. A spider has eight legs instead of six and no antennae (feelers) or wings. Spiders have poor eyesight and must rely on their sense of touch. All spiders are helpful to people because they eat insects, such as flies, mosquitoes, and gnats. This activity will help children discover some of the differences between spiders and insects.

## MATERIALS

- **PICTURES OF SPIDERS AND INSECTS**
- **LIVING, DEAD, OR PLASTIC SPIDER**
- **LIVING, DEAD, OR PLASTIC INSECT**
- **MAGNIFYING GLASS**
- **TWEEZERS**
- **SCIENCE JOURNALS**
- **COLORED PENCILS**

## ACTIVITY

1. Display all of the materials in the science area during self-selected activity time.

2. As children approach the science area, suggest they look for similarities and differences between the spider and insect. Ask them open-ended questions like these:
   - How do spiders move?
   - What do spiders eat?
   - Where do spiders live?
   - How do spiders help people?
   - How are spiders like insects? How are they different?

   Encourage children to document their observations in their science journals.

## ADDITIONAL ACTIVITIES

- **ART**—Provide pieces of Styrofoam, egg cartons that have been cut apart, pipe cleaners, pompoms, glue, scissors, and other materials for children to create spiders or insects during self-selected activity time. Display pictures of spiders and insects near the art center so children have a point of reference.

## Ideas from Sherri's Classroom

This activity can easily stem from a teachable moment any time of the year. One winter, my preschoolers discovered a spider in the hallway leading to our bathrooms. The bug box was retrieved and the spider spent the morning on our science table among magnifying glasses, field guides, science journals, and flashlights. The children tried to identify it; discussed among themselves how they knew it was a spider rather than an insect; talked about what it ate, where its web was located, and why it was inside. Toward the end of the school day, one of the children suggested that the spider be released outside. Someone else remarked that it was cold outside and no spiders were around. After much discussion, the children decided to release the spider back into the hallway where it had chosen to spend the winter. (As the teacher, I am familiar with the two types of venomous spiders found in my area—brown recluse and black widow. I knew the spider was neither of these so felt comfortable allowing the children to make this decision.) Although there was some screaming and reluctance about sharing our school habitat, the children watched the spider closely as they released it. Of course the spider didn't move much until the noise died down and many of the children had gone home. The next week our hallway was actively investigated to find our new winter resident.

- **FIELD TRIP**—Tell children and parents how to go spider-sniffing at night. Holding a flashlight next to your eyes, scan the tree line. Spiders' eyes have a reflection that will show in the flashlight that only the person holding the light will be able to see. Follow the reflection to the spider.

- **MANIPULATIVE**—Display a variety of animal pictures. On a chart, create columns labeled "No Legs," "Two Legs," "Four Legs," "Six Legs," and "Eight Legs." During choice time, encourage children to sort the animal pictures by the number of legs the animal has.

- **NUTRITION**—Serve large and miniature marshmallows and pretzel sticks for snack. Encourage children to create spiders and insects with them. Discuss the body parts of spiders and insects as they work.

- **STORY**—Read *The Web in the Grass* by Berniece Freschet (see appendix A on p. 188). This book illustrates how a spider spins a web to capture its food and what it eats. It also describes the spider's enemies and how the spider reproduces.

# Spiderweb Toss ▨ Large Motor

## DID YOU KNOW?

The spider's web has a sticky surface that holds an insect entangled until the spider returns to kill it. Once the insect is dead, the spider wraps it in a silken sac for food to be eaten later. This activity will enable children to explore how spiders capture their food.

## MATERIALS

- CLEAR CON-TACT PAPER CUT INTO LONG STRIPS
- LIGHTWEIGHT OBJECTS TO REPRESENT INSECTS (SUCH AS POM-POMS, STYROFOAM PACKING, OR BITS OF PIPE CLEANERS)

## ACTIVITY

1. Use the strips of Con-Tact paper to make a spiderweb in a low corner of the classroom. Be sure to position the Con-Tact paper with the sticky side out.

2. During self-selected time, encourage children to try throwing various "bugs" into the web to see what the spider captures for its lunch. Ask children open-ended questions like these:
   - How do real spiders make their webs sticky?
   - What kind of insects usually get caught in spiderwebs?
   - What keeps the spider from sticking to the web?

## ADDITIONAL ACTIVITIES

- **ART**—Provide white chalk and black paper for children to draw spiderwebs during self-selected activity time. Display pictures of webs in the art area to provide inspiration and reference.
- **FIELD TRIP**—Look for spiderwebs on a field trip to a natural area. Spiders usually make their webs between branches of bushes, trees, or other plants. Orb webs can be coated with dark colored, nonlacquer spray paint and transferred onto white drawing paper. Talk about the many different types of spiderwebs seen on the trip. Be sure to take science journals along for children to record and sketch their observations.
- **MANIPULATIVE**—Make a spider lotto game for children to match pictures of spiders with pictures of the type of web they weave. This is especially meaningful when you include pictures of common spiders indigenous to your area.
- **NUTRITION**—Make "Peanut Butter Playdough" (mix 1 cup peanut butter, ¼ cup honey, and enough dry milk to prevent stickiness). Make playdough insects for lunch and encourage children to pretend they are spiders.
- **OUTSIDE**—Adopt a spider on or near your play yard. Observe the spider and its habits on a regular basis. Encourage children to record observations in their science journals.
- **SCIENCE**—Prepare a collection of nonvenomous spiders to display in the science area for children to examine.
- **STORY**—Read *Spider's Web* by Christine Back (see appendix A on p. 187). This nonfiction book uses photographs to describe how a garden spider spins a web and catches food.

# The Lights Go On ▥ Music

## DID YOU KNOW?

Fireflies, also called lightning bugs, are beetles, and they have a unique organ in the body that flashes a light to attract its mate. The females, resting on the ground, flash their pattern to attract a male of her own species who answers with a similar flash and flies to her. This activity will help children develop an understanding of this familiar creature.

## MATERIALS

- **PICTURES OF FIREFLIES**

## ACTIVITY

1. Teach children the following song:

First they're dark and then they're light, The light-ning bugs are out to-night, They sleep all day, they come out late, Then flash their lights to find a mate.

2. After children have learned the song, ask them open-ended questions like these:
   - How do fireflies know when to light their lights?
   - What do you think happens to fireflies in the winter?
   - Where do fireflies go in the daytime?

Fireflies are common throughout the country. The magical quality of their flashing lights is a fond memory of many childhoods. Each species of lightning bug has a unique flash pattern that allows experts to identify them by their flash pattern alone. Scientists use the light organs of fireflies in biochemical analyses; however, they are gathered from wild firefly populations because people haven't been able to raise them in capitivity.

## ADDITIONAL ACTIVITIES

- **FIELD TRIP**—Plan an evening field trip to capture fireflies. Be sure to release the fireflies after children are finished observing them.
- **OUTSIDE**—Observe the pattern of firefly blinking and try to fool the fireflies by using a flashlight.
- **SCIENCE**—Place a jar of fireflies in a pan of cold water and another in a pan of warm water. Observe the differences in the amount of light the fireflies produce. Make science journals available for children to document their observations.
- **SCIENCE**—Capture other beetles and encourage children to compare them to fireflies.
- **STORY**—Read *Fireflies!* by Julie Brinckloe (see appendix A on p. 182). In this book, a young boy proudly shows off his newly acquired piece of moonlight—a jar of fireflies—but realizes he must set them free as their lights begin to fade.

# Ant Café ▧ Outside

## DID YOU KNOW?

There are more ants than any other creature on earth. Ants are social insects and live in colonies. When an ant discovers a good meal, it rushes back to the colony, leaving a scent trail back to the meal. Hundreds of ants are then able to follow the trail back to the meal. This activity will encourage children to discover ants and their habits.

## MATERIALS

- **WHITE PAPER PLATES**
- **FOOD FOR THE ANTS: SUGAR, HONEY, LEAVES, GRASS, BREAD CRUMBS, AND ANY OTHER FOOD CHILDREN THINK ANTS MIGHT EAT**
- **CHART PAPER AND MARKER**
- **SCIENCE JOURNALS**

## ACTIVITY

1. Present this activity to younger children during choice time so they can move in and out of it. You could present this as a group activity to older children but care should be taken to keep waiting to a minimum.

2. Encourage children to select foods they think ants might like and place them on different parts of the paper plate.

3. Place the paper plate in the play yard away from the building.

4. Experiment with different locations to see if habitat affects the number of ants feeding at the "café." For example, place one plate in the grass and another on a hard-packed dirt area. Ask children open-ended questions like these:
   - Which food do you think the ants will like best?
   - How do the ants know where the food is?
   - Why do the ants stay in a line?
   - What would happen if you rubbed your finger across the line?
   - What do ants eat when people aren't around?

5. Record the children's observations about the ants who come to dine at the café. Pictures may be taken to help children revisit the experience in the classroom. Encourage children to use their science journals to document their observations and thoughts.

## ADDITIONAL ACTIVITIES

- **GROUP**—Invite an entomologist (either a professional or a 4-H student) to bring an insect collection to class.
- **OUTSIDE**—Go on an insect scavenger hunt. Ask children to find a place where an insect has lived, something an insect has nibbled on, signs of where an insect has been, and so on.
- **STORY**—Read *Ant Cities* by Arthur Dorros (see appendix A on p. 183). This nonfiction book describes ants and their habits.

# Roly-Poly Paradise ▧ Science

## DID YOU KNOW?

The roly-poly, or pill bug, is one of the most common land-dwelling crustaceans. This small armadillo-like creature has seven pairs of legs (unlike insects) and rolls into a ball, like a hard, black pill when frightened. This experience will allow children to explore some of the habits of the roly-poly.

## MATERIALS

- **PILL BUGS (THESE CAN BE FOUND IN LEAF LITTER OR NEAR DAMP, ROTTING LOGS)**
- **CONTAINERS**
- **MAGNIFYING GLASSES**
- **SCIENCE JOURNALS**

## ACTIVITY

1. During self-selected activity time, provide pill bugs for children to investigate.

2. Encourage children to use the magnifying glasses for observing the pill bugs. Ask them open-ended questions like these:
   - What does a roly-poly need to live? How does a roly-poly move?
   - How do you know a roly-poly isn't an insect?
   - Why do you think the roly-poly makes a ball?
   - How does it know when to unroll?

   Have science journals available for children to document their observations.

3. Before the end of the day, challenge children to find suitable homes for their pill bugs and carefully release them from their containers to their new homes.

## IDEAS FROM SHERRI'S CLASSROOM

Next to our sandbox, there is a large log the children move and explore under on a routine basis. Roly-polies and earthworms are always there, and the children frequently discover other inhabitants as well.

They are always careful to return the log to the original location and position when they are finished exploring so that the habitat is restored.

After our experience with the crayfish (see the introduction to this book on p. 3), the children were amazed to find in their research that roly-polies are also crustaceans. They immediately tried to find similarities between the two creatures.

## ADDITIONAL ACTIVITIES

- **ART**—Sketch roly-polies.
- **FIELD TRIP**—Go on a roly-poly expedition looking for places where pill bugs live.
- **SCIENCE**—Set up experiments to determine roly-poly habitat preference. For example, place the roly-poly in a container with dry soil at one end and wet soil at the other. Observe where the roly-poly goes. Other experiments might include preferences for light versus dark, open areas versus enclosed spaces, and so on.

# Trash Pickup ▨ Field Trip

## DID YOU KNOW?

Littering is more than just cluttering the ground with paper, cans, and bottles. Littering is unwise use of our natural resources, many of which are limited. Litter also affects our environment—it changes the environment and makes the outdoors less pleasant for people and unsafe for wildlife. This activity will help children better understand the effects of littering and how they can help to preserve the land.

## MATERIALS

- **LITTERED AREA (SUCH AS A PARK, DITCH ALONG A ROAD, AND SO ON)**
- **GROCERY BAGS (ONE PER CHILD)**
- **DISPOSABLE OR WASHABLE WORK GLOVES**

## ACTIVITY

1. Distribute a grocery bag to each child and explain that they will be going on a litter walk.

2. Explain the boundaries of the area to be cleaned, and challenge children to pick up all the litter they find. Be sure the children wear gloves for protection.

3. Talk with children about litter that may be unsanitary or unsafe for them to pick up. It will require a grown-up with rubber gloves to pick up. Some unsafe or unsanitary items include dirty diapers, cigarette butts, and broken glass.

4. As children finish collecting trash, discuss changes in the area. Be sure to talk about the environmental impact that cleaning up the area has provided. Ask children open-ended questions like these:
   - Where did all this trash come from?
   - What shall we do with the trash we collected?
   - What would happen if no one ever picked up the trash?
   - How can we make sure this doesn't become littered again?
   - How does this litter affect habitat for creatures and plants who live in the littered area?

## IDEAS FROM SHERRI'S CLASSROOM

Tires, when placed in a landfill, never decompose; however, they can be recycled in a number of ways. One of these possibilities, perhaps new to you and your children, actually uses old tires to create habitat. Tires can be cut in half and sunk into a lake to create instant nesting habitat for catfish. The idea that all trash is either "bad" or "good" is a difficult concept for both children and adults. Considering new ways of recycling our trash is an important step in "wise use" of natural resources.

## ADDITIONAL ACTIVITIES

- **ART**—During self-selected activity time, provide children with paper bags, markers, crayons, glue, construction paper, and other materials for decorating litterbags for trash collection at home or in the car.

- **BLOCK**—During self-selected activity time add dump trucks and cranes to the play area for sorting and hauling trash. Use real trash or Styrofoam pieces to represent trash.

- **BULLETIN BOARD**—Have children create a beautiful outdoor scene, then tack litter all over it. Discuss how the trash effects both the people who live in the area as well as the wildlife.

- **FIELD TRIP**—Visit a recycling plant or landfill.

- **GROUP**—Encourage children to clean up their own messes after routine activities such as art, snack, washing hands, or blowing noses.

- **GROUP**—Model wise use of paper for children by using both sides and encouraging them to do the same.

- **PRETEND PLAY**—During self-selected activity time provide recycling bins with signs designating paper, aluminum, and plastic. Challenge children to sort clean trash into the proper bin. Include an aluminum-can smasher for children to compact trash.

- **PRETEND PLAY**—Provide coveralls and heavy gloves for children to dress up like garbage collectors and pick up trash. Create a garbage truck out of a large box and invite children to "drive around" and collect the garbage during self-selected activity time.

- **STORY**—Read *The Wump World* by Bill Peet (see appendix A on p. 185). This is the story of how litter and pollution changed the Wump World.

# What Is Air?  Science

## DID YOU KNOW?

All living things are immersed in air. You cannot see or touch air, but you know it's always there. This experience will help children become more aware of air.

## MATERIALS

- **OBJECTS FOR MOVING WITH AIR (SUCH AS FEATHERS, PING-PONG BALLS, ROCKS, SHELLS, AND LEAVES)**
- **STRAWS**
- **SCIENCE JOURNALS**

## ACTIVITY

1. Gather materials together in a place where children have room to move along the floor.

2. During self-selected activity time, provide drinking straws for children in this play area. Challenge them to blow through the straws and to try and move the various objects you have placed in the area.

3. Ask children open-ended questions like these:
   - How do you know air is there when you can't see or feel it?
   - Where does air come from?
   - How does air move things?
   - How do you know it is air moving the object?

   Provide science journals for children to document their observations and discoveries.

## Additional Activities

- **Art**—During self-selected activity time, add strong coloring to a bubble solution. Have children create bubble prints by using straws to blow bubbles onto absorbent paper. Discuss how air makes bubbles and what happens to the air when the bubbles pop.
- **Outside**—Go outside on a windy day and "feel" the air. Watch for signs of air, such as blowing leaves, moving clouds, and so on.
- **Story**—Read *The Four Elements: Air* by Maria Ruis and J. M. Parramon (see appendix A on p. 177). This book describes air and its uses in simple terms.

# What's in the Air?  Science

## DID YOU KNOW?

Although air pollution is generally thought to be primarily in the cities or near roadways, particles of dust and other pollutants can be found in the air everywhere. This activity will help children detect dust and other pollutants in the air.

## MATERIALS

- **INDEX CARDS**
- **PETROLEUM JELLY**
- **PLASTIC KNIVES OR CRAFT STICKS**
- **CHART PAPER AND MARKER**
- **MAGNIFYING GLASSES**
- **SCIENCE JOURNALS**
- **AUDIO RECORDER AND BLANK CASSETTE (OPTIONAL)**

## ACTIVITY

1. During group time, talk with children about the experiment. Discuss the kinds of things that might be in the air. As a group, decide how many cards should be prepared for the experiment and where the cards should be placed in the environment.

2. During self-selected activity time, place the cards, petroleum jelly, and plastic knives or craft sticks in the science area. Encourage children to spread a thin layer of the petroleum jelly on the cards.

3. Allow children to place the cards throughout the environment, both indoors and outdoors.

4. Add columns to the chart paper that correspond with the number of cards placed in the environment. At the top of each column, describe where each card was placed.

5. After several hours, check the cards and discuss the material found on them. Ask open-ended questions like these:
   - Where did this material come from?
   - Why do some of the cards have more material on them than others?

- What happens when we breathe these particles?
- What happens when animals breathe them?
- Why can't we see these particles when we breathe them?
- Where do you think the particles come from?

Be sure to provide magnifying glasses so children can see the fine particles that have collected. Have science journals available so children can document their discoveries.

6. Record their observations on the chart and discuss their discoveries. The discussion can be taped and transcribed to aid in assessing their understanding and determining future study.

## ADDITIONAL ACTIVITIES

- **FIELD TRIP**—On a dry day, gather leaves from a roadside ditch (this should be done with close adult supervision). Rub leaves with a damp, white tissue and examine the tissues. Compare leaves taken from several different ditches with those taken from a park or the play yard. Mark samples according to location to help provide a clearer understanding of the results of the experiment.
- **GROUP**—Discuss other forms of pollution, such as noise, land, water, and so on.
- **PRETEND PLAY**—Experiment with placing oil and water in the sensory table or a large tub. Add approximately ½ cup cooking oil to the water. Allow children to play in the water as usual, with various plastic animals. During cleanup, encourage children to wash the animals. Talk about the way the water felt with the oil added and discuss the difficulty in cleaning up the mess. Relate this to water pollution.

- **SCIENCE**—Repeat the experiment in other places the children frequent. Compare the results.
- **STORY**—Read *A River Ran Wild* by Lynne Cherry (see appendix A on p. 185). This book tells the environmental history of the Nashua River.

# Aquatic Life ❀ Field Trip

## DID YOU KNOW?

Wildlife is often hard to see in the natural environment. However, a careful observer can often spot evidence of wildlife inhabitants in or around a body of water. This activity will enable children to see aquatic life in its natural habitat.

## MATERIALS

- **BUTTERFLY OR AQUARIUM NETS**
- **SEVERAL 3-POUND COFFEE CANS**
- **SEVERAL NUTS AND BOLTS**
- **SEVERAL 2- TO 3-FOOT FLAT STICKS**
- **PAPER MILK CARTONS (FOR VIEWERS)**
- **PLASTIC WRAP**
- **DUCT TAPE**
- **BUCKETS**
- **DISHPAN**
- **CLIPBOARDS**
- **PAPER**
- **COLORED PENCILS**
- **MAGNIFYING GLASSES**
- **GARBAGE BAG**
- **CAMERA (OPTIONAL)**
- **POND, LAKE, OR STREAM**

## ACTIVITY

1. Prepare the nets, sieves, and viewers prior to going on the field trip. Sieves can be made from the cans and sticks. Punch holes in the bottom of the cans, and bolt one stick or handle to each side. Viewers should be made by removing and covering the bottom of a milk carton with plastic wrap. Plastic wrap should be secured with duct tape.

2. Prior to leaving on the field trip, talk with children about what they might see in and near the pond or stream. Brainstorm a short list of rules and cautions for the children to follow while on the field trip. Be sure to recruit extra adults to provide adequate supervision.

3. During the field trip, use the nets and sieves to collect samples from the water. Place samples in the tub, along with pond or stream water, for children to observe using their viewers and magnifying glasses.

4. Walk around the area looking for other animal signs, such as tracks, nibbled plants, smashed down grass, and so on. Ask children open-ended questions like these:
   - What do these animals and plants need to survive?

## IDEAS FROM SHERRI'S CLASSROOM

Visiting a body of water on a routine basis encourages children to explore the inner dependency of the ecosystems involved, as well as discovering the seasonal changes in the habitat. I have found that visiting the same area several times a year also helps the children feel responsible for the area and its inhabitants. Although this type of field trip requres extra adult supervision, it is well worth the effort.

---

- • What do you notice about these animals that helps them in their habitats?
- • How can you tell where an animal has been?

5. As children explore the aquatic life in the area, provide clipboards, paper, and colored pencils for them to sketch the things they observe. Adults should sketch along with the children, providing role models (not art models) to help children focus.

6. If children notice trash in the area, use the garbage bag to collect and properly dispose of it.

7. The camera may be used to photograph aquatic life that is collected, tracks, and other evidence of wildlife inhabitants, allowing children to revisit the experience back in the classroom.

8. Be sure to release all collected aquatic life before leaving the area.

## ADDITIONAL ACTIVITIES

- • **ART**—During self-selected activity time, encourage children to make pictures of aquatic life by first drawing and coloring fish and other aquatic life with crayons, and then painting a blue watercolor wash over the drawing.
- • **LARGE MOTOR**—Challenge children to move like various animals that live in water.
- • **MANIPULATIVE**—Provide two shallow boxes (one colored like soil and the other colored like water) and pictures of various animals during self-selected activity time. Have children place pictures of land animals in the "soil" box and pictures of aquatic animals in the "water" box. Be sure to provide pictures of animals that might live in both (for example, frogs, turtles, or snakes).
- • **PRETEND PLAY**—During self-selected activity time provide sticks, mud, and leaves at the water table for children to try building beaver dams.
- • **SCIENCE**—Prepare an underwater "feely box" using smooth rocks, fish scales, plastic worms, muscle shells, and other materials for children to use during self-selected activity time.
- • **SCIENCE**—Bring water samples from a creek, stream, pond, or lake into the classroom for children to explore. Look for living creatures. Talk about the color of the water as well as the creatures living in it. Provide science journals and magnifying glasses for children to investigate and document their findings.
- • **STORY**—Read *Pond Year* by Kathryn Lasky (see appendix A on p. 179). This beautifully illustrated book shows two young girls enjoying and exploring a pond throughout the year.
- • **WRITING**—Use the children's sketches and photographs to create a class book about the experience. Include a pocket and card so it can be checked out and shared at home. Create an audio recording of children reading their entries, and allow for it to be checked out as a "book on tape."

# Joe's Choice ▓ Story

## DID YOU KNOW?

Conservation means wise use of natural resources. There are three levels of conservation: preservation, restoration, and management. All three are conservation practices, but the situation determines which practice should be used. This activity will introduce the idea of conservation to the children and demonstrate the three levels of conservation.

## MATERIALS

- **SUMMER PATTERN 1 (SEE APPENDIX D, P. 234)**
- **FELT CHARACTERS**
- **FLANNELBOARD**

## ACTIVITY

1. Read the following story to children, placing the felt characters on the flannelboard at the appropriate time. You should be able to readily identify stanzas that deal with restoration, preservation, and management.

2. After reading the story, ask children open-ended questions like these:
   - What would have happened to the lake if Joe hadn't practiced conservation?
   - What else could Joe have done to solve his fishing problem?

Joe caught a bunch of fish
from his fishing lake.
He took them home for dinner
and he made a big fish cake.

He ate the cake. YUM! It was good.
And catching fish was fun.
So he went to catch some more,
just as fast as he could run.

But the fishing lake was fished out.
He'd caught every single one.
No more fish to catch and eat.
What can be done?

Joe then bought some fish.
He put them in his lake.

Then he went and caught the fish
and made another cake.

Again his lake was fished out.
He said "What can I do?
I fished until there's no more fish.
Boo-hoo-hoo!"

So, once again he bought some fish.
But, before he did he thought,
"I'll put these fish into my lake
and then I'll sit and watch."

Joe sat and watched his lake
to see how his fish would do.
But every time the fish jumped,
Joe's stomach jumped too!

"I like to watch the fish," said Joe,
"but I like to eat them too.
How can I catch a fish
and leave the lake a few?"

So Joe began to fish again
but he only kept a few.
He'd let the smaller fishes go
and just keep one or two.

The smaller fish, they grew and grew.
The bigger ones, Joe ate.
Joe's lake was full of fish
and the fishing there was great.

—John Griffin

## ADDITIONAL ACTIVITIES

- **OUTSIDE**—Provide shovels and a dirt area for children to dig for fishing worms during self-selected activity time.
- **WRITING**—Provide blank books and colored pencils for children to write and illustrate their own fishing stories.

# Fishing Permits ⬛ Art

## DID YOU KNOW?

In most states, everyone between the ages of sixteen and sixty-four who fishes in state or federal water must have a fishing permit. These permits help the state department's wildlife agency keep track of the number of people fishing. The revenue from these permits helps pay for the management of state lakes and streams. This activity will help children to be aware of the need for fishing permits as well as introduce them to the role of conservation agents or game wardens.

## MATERIALS

- **WASHABLE STAMP PAD**
- **CRAYONS**
- **INDEX CARDS**
- **PENCILS**
- **INSTANT CAMERA OR PHOTO OF EACH CHILD (OPTIONAL)**
- **SCISSORS**
- **GLUE STICKS**
- **OFFICIAL FISHING PERMIT (OBTAIN ONE FROM YOUR STATE DEPARTMENT IF POSSIBLE)**

## ACTIVITY

1. Place materials in the art area during self-selected activity time. If using the instant camera, photograph groups of children and cut the photos apart.

2. As children approach the art area, show the real fishing permit. Talk with them about the need for fishing permits and about the information included on them.

3. Encourage children to create their own fishing permits. Discuss what they might include on their permit. Ask children open-ended questions like these:
   - Why do people need fishing permits?
   - What information needs to be on your fishing permit? (Responses might include their picture; a drawing of themselves; their fingerprint; or their name, address, and other important information.)
   - What do conservation agents have to do with fishing permits?

- What else do conservation agents do besides enforce fishing regulations?

4. Talk with children about the job of conservation agents or game wardens. They are like police officers for wildlife. Conservation agents check fishing permits and enforce hunting and fishing regulations.

## ADDITIONAL ACTIVITIES

- **ART**—Repeat this activity for hunting permits during deer, turkey, quail, squirrel, or other hunting seasons that occur in your state.
- **FIELD TRIP**—Visit a conservation agent at a state lake or wildlife area.
- **PRETEND PLAY**—Provide conservation agent or game warden dress-up clothes, such as hats, shirts with patches on the sleeves, badges, ticket books, pencils, binoculars, radios, and so on. Provide billfolds and purses to keep the fishing permits in. Encourage children to use these during self-selected activity time.

# Let's Go Fishing ▨ Pretend Play

## DID YOU KNOW?

Fishing is a sport enjoyed by many people throughout the country. There are many different kinds of fish and other wildlife found in creeks, streams, lakes, and ponds—and they are fun to catch. However, the state wildlife code determines which fish or wildlife and how many of each may be kept by people who are fishing. This activity will help children explore some of the rules concerning fishing, while engaging them in a fun fishing activity.

## MATERIALS

- **SUMMER PATTERN 2 (SEE APPENDIX D, PP. 235–236)**
- **STURDY STICKS**
- **HEAVY STRING**
- **MAGNETS**
- **PAPER CLIPS**
- **RULERS**
- **LAMINATED FISH, TURTLES, CRAYFISH, SNAKES, FROGS, AND TRASH**
- **CHILD-SIZE PERSONAL FLOATATION DEVICES (OPTIONAL)**
- **BLOCKS FOR CREATING A BOAT DOCK**
- **BOATS MADE FROM CARDBOARD BOXES**

## ACTIVITY

1. Have the children create fishing poles by attaching a strong magnet to an arm's length of string and then tying it to a stick. Place a paper clip on the end or mouth of each fish, turtle, crayfish, snake, or other animal.

2. Use the blocks to build a fishing dock in the block area. If your area is large enough, boats made from cardboard boxes can be added as well.

3. During self-selected activity time, encourage children to experiment with fishing. Have them wear life jackets, and talk about the importance of boat safety while fishing. Ask children open-ended questions like these:
   - What happens when all the fish are caught?
   - How can we make sure there are enough fish for everyone to catch some?
   - What should you do when you catch something other than fish?
   - What could you use the rulers for?
   - What happens if you fall in the water?

This activity is a favorite in my classroom. It always provides an excellent opportunity to consider and apply the three levels of conservation. Generally when I make the activity available, the first children to participate catch all of the fish, which causes a problem for the remaining children who want to fish. Usually a class meeting is called where the children discuss how to manage the problem. They always discuss putting the activity away (preservation), limiting the number of fish each child can catch (management), and making more fish or throwing the fish back (restoration). Children are learning about conservation ethic that is directly applied to their experience.

## ADDITIONAL ACTIVITIES

- **FIELD TRIP**—Take children to a nearby lake to go fishing.
- **LARGE MOTOR**—Show children how to play "Fish and Worms." Divide the class into two groups: half the children are "fish" and the other half are "worms." The fish chase the worms, trying to catch their lunch.
- **NUTRITION**—Serve sardines for snack as a tasting experience. Talk about what kind of fish they are and where they live. Try other kinds of fish to compare the tastes.
- **MANIPULATIVE**—During self-selected activity time, display a tackle box that contains different kinds of fishing tackle (with hooks removed) for children to sort and classify.
- **PRETEND PLAY**—During self-selected activity time place empty containers from cornmeal, oil, butter, and flour, as well as skillets, pots, and pans in the pretend-play area for children to pretend to cook fish as they catch them.
- **PRETEND PLAY**—In most parts of the country, frogging is as popular a sport as fishing. Set up a frogging experience for children during self-selected activity time. Cut out and laminate pictures of frogs and attach a small strip magnet to each picture. Create frog gigs by hammering a nail into one end of a 1- to 1½-inch dowel rod. Use blocks to build docks, and remember to include the boats and life jackets. Be sure to include materials for cooking the frog legs as well.
- **OUTSIDE**—Place a flat-bottomed boat, oars, and flotation devices on the play yard for children to pretend to fish and participate in various water sports.
- **OUTSIDE**—During self-selected activity time place small fishing rods and reels as well as hula hoops in the play yard. Encourage children to practice casting into the hula-hoop ponds.
- **SCIENCE**—For a brief period of time, bring live fish into the classroom for children to observe. Be sure to have sketchboards and science journals available for children to document their observations.

# Frogs and Toads ▩ Story

## DID YOU KNOW?

All amphibians hatch from eggs and go through a metamorphosis. This story illustrates that process while clarifying some of the differences between frogs and toads. In addition, children will also hear more about predator and prey relationships and some ways frogs and toads avoid predators.

## MATERIALS

- **SUMMER PATTERN 3 (SEE APPENDIX D, PP. 237–238)**
- **FLANNELBOARD**
- **FELT CHARACTERS**

## ACTIVITY

1. Read the following story to children, and place the felt pieces on the flannelboard at the appropriate times in the story:

2. After reading the story, ask children open-ended questions like these:
   - Where do you think you might find frog or toad eggs?
   - What do tadpoles need to grow into frogs or toads?
   - What would have happened to the animals that ate the eggs and tadpoles if there hadn't been any eggs or tadpoles?

## ADDITIONAL ACTIVITIES

- **ART**—Provide green and brown fingerpaint during self-selected activity time. Add a small amount of sand to the brown. Discuss differences in texture and how this relates to frog and toad skin.
- **ART**—During self-selected activity time, offer colored chalk for children to create pond, lake, and stream art. Provide scissors for them to cut out frogs and toads from their dry fingerpaintings (see above) and glue onto their chalk background. These could lead to a discussion of habitat and the needs of frogs and toads for sur-

In a very shallow pond
where you could walk and wade,
in among the weeds and grass
were gobs of gooey eggs.

Small black eggs in slimy jelly
made in long thin lines,
other eggs in big huge hunks
wrapped up in globs of slime.

Along came a snapping turtle—
he ate a slimy batch.
But many eggs were left untouched;
soon they began to hatch.

From the eggs came tadpoles.
They wiggled when they swam.
Along came a hungry fish
and ate a bunch of them.

But more were left and as time passed,
they began to grow.
Soon they started sprouting arms,
legs, and feet, and toes.

The tadpoles' tails began to shrink.
They used their legs to swim.
Along came a hungry heron.
He ate some more of them.

They swam up from the bottom
to sunlight shining there.
They poked their heads out of the
pond
and started breathing air.

Up out of the water
they crawled toward the shore.
Along came a hungry snake
and gobbled up some more.

They jumped out of the water.
They hopped and jumped on logs.
The tadpoles from the farmpond
had changed to toads and frogs.

Out flashed their sticky tongues.
They gobbled bugs and flies
and whatever they could swallow
if it happened to crawl by.

Some had skin smooth and sleek,
their hops were strong and long.
They could jump into the water,
if they thought something was wrong.

They were better swimmers.
They would sunbathe on a log.
They were big and quick and strong,
we always call them frogs.

Then there were the others.
They were fat and short.
Their skin was rough and covered
with many bumpy warts.

They were small and stocky.
Their hops were short and slowed.
They weren't good at swimming.
They are known as toads.

Along came a coyote
looking for a meal.
He jumped into the toads and frogs.
He gave a hungry squeal.

The frogs were fast. They jumped away
as fast as they could go.
But the coyote snapped a toad
right up.
The toad was much too slow.

Breakfast for the coyote.
He started to bite down.
Bleck! The toad, it tasted bad.
He spit it on the ground.

The warts that cover toad skin
make them horrible to eat.
The coyote had to go and search
for better tasting meat.

—John Griffin

vival. Children could then be challenged to
include other elements in their habitat artwork.

- **LARGE MOTOR**—Encourage children to act out
  metamorphosis of a frog or toad. Begin as an
  egg, develop into a tadpole, and finally become a
  frog or toad.
- **LARGE MOTOR**—During self-selected activity
  time, play leap frog. Discuss how far children can
  hop. Compare their leg lengths with the distance
  of their hops.

Frogs and toads are amphibians that are easily observed throughout the country. This song works well after a frog or toad has been discovered in the play yard or someone has brought one in to share.

- **MANIPULATIVE**—Laminate pictures of frogs and toads indigenous to your area. Place in the manipulative area during self-selected time and encourage children to sort them by type.

- **MUSIC**—Teach children the following song to the tune of "Frère Jacques":

  Frogs and toads.
  Frogs and toads.
  Toads and frogs.
  Toads and frogs.
  Frogs have long legs. (*Place hands far apart.*)
  Toads have short legs. (*Place hands close together.*)
  Frogs go jump. (*Make hands jump high.*)
  Toads go hop. (*Hop hands just a little.*)

- **NUTRITION**—Make two kinds of treats for snack: green gelatin jigglers (see recipe on box) and rice krispy treats with cocoa-flavored cereal. Cut both kinds of treats with a cookie cutter shaped like a frog and serve for snack. As the children are eating, discuss the differences between frog skin (represented by the gelatin jigglers) and toad skin (represented by the cocoa krispy cereal treats).

- **SCIENCE**—Bring in a frog and a toad, and place them in the science area for children to observe during self-selected activity time. Provide science journals for children to document their observations. Encourage children to design a habitat for each. How would they be alike? How would they be different? Be sure to return the frog and toad back to their original habitat after children finish observing them.

- **STORY**—Read *All about Frogs* by Jim Arnosky (see appendix A on p. 177). This nonfiction picture book illustrates differences between frogs and toads as well as provides information about life cycles, calls, and predators.

# You Can't Find Me! ▧ Outside

## DID YOU KNOW?

One of the many ways animals protect themselves from predators is with camouflage. This activity will allow children to explore the concept of animal camouflage and its importance.

## MATERIALS

- **SMALL STRIPS OF NATURAL-COLORED YARN OR THIN RIBBON, OR COLORED TOOTHPICKS**

## ACTIVITY

1. Before introducing the activity to children, scatter the pieces of yarn or ribbon in a grassy area.

2. Take children outside and explain that they are going to be hungry birds looking for a tasty worm or caterpillar represented by the material selected (yarn or ribbon).

3. Encourage children to find as many "worms" or "caterpillars" as possible within a specified time.

4. Give children about five minutes, then have them bring their "lunches" and talk about the number of worms and caterpillars caught. Ask open-ended questions like these:
   - Which worms or caterpillars were the easiest to find? Why?
   - Which worms or caterpillars were the hardest to find? Why?
   - How do you think color helps animals?

## ADDITIONAL ACTIVITIES

- **ART**—During self-selected activity time, put out paint and paper of the same color at the easel. Discuss the painting results.
- **FIELD TRIP**—Take children on a hike in a natural area. Hide several pictures of animals along the trail. Challenge children to discover the animals. Discuss which animals were easiest to find, which were most difficult to find, and why.
- **MANIPULATIVE**—Cover three boxes and place a feather on the end of one, a scrap of fur on another, and a scrap of smooth, scale-like material on the third. During self-selected activity time, encourage children to sort pictures of animals by skin type. Discuss how the skin type helps camouflage the animal in its habitat.
- **STORY**—Read *Can You Find Me? A Book about Animal Camouflage* by Jennifer Dewey (see appendix A on p. 179). This nonfiction book describes how animals use camouflage to locate food and avoid enemies.

# Foxy Predators ▨ Large Motor

## DID YOU KNOW?

Animals that eat and hunt other animals are called *predators*. The animals they hunt are called *prey*. This activity will enable children to explore the relationship between predators and prey.

## MATERIALS

- **BROWN CONSTRUCTION PAPER (RABBIT)**
- **RED CONSTRUCTION PAPER (FOX)**
- **SCISSORS**
- **STAPLER**

## ACTIVITY

1. Create headbands for children by cutting long strips of red and brown construction paper. (Two pieces may need to be attached together to fit a child's head.)

2. Cut fox and rabbit ears and attach to the appropriate colored headband.

3. Gather children together and explain that you are going to play a hunting game. (This game works best in a large, outdoor space with lots of hiding places.) The foxes, or predators, are very hungry and are going hunting for their prey, the rabbits. But the rabbits don't want to be eaten.

4. Give the rabbit headbands to the rabbits and the fox headbands to the foxes.

5. The foxes should cover their eyes and slowly count to ten while the rabbits go and hide.

6. After reaching ten, the foxes search for and try to sneak up on the rabbits. Once tagged, the prey have been eaten and should sit on the ground and wait until all the prey have been eaten or until it is announced that the game is over.

7. Discuss the relationship between predators and prey. Ask children open-ended questions like these:
   - How can the prey keep from being eaten?

## IDEAS FROM SHERRI'S CLASSROOM

Predator-and-prey relationships are an important part of nature. Prey provide food for predators but predators also help keep prey populations healthy. Without predators, the prey poulation would grow too large for its food source and disease would spread more easily. The prey population would eventually end up starving to death. The number of prey available helps control the number of predators. Without prey, predators would starve to death. The delicate balance between the two is vital to a healthy ecosystem.

One well-known example of people tampering with this balance took place in Arizona during the early part of the twentieth century. In 1907, under the leadership of our first conservationist president, Teddy Roosevelt, a bounty was placed on all of the natural predators, such as mountain lions and wolves, of mule deer on the Kaibab Plateau, located on the north rim of the Grand Canyon. It was hoped that this would increase and protect the deer herd. As a result of this bounty, the predator population reduced dramatically and in a short span of fifteen years, the deer population devoured all the available vegetation.

During the winter of 1924, the ecosystem inevitably crashed and an estimated 60 percent of the deer population died. Left in its wake was a diminished deer population and a depleted ecosystem.

Adults often think children should be sheltered from predator-and-prey relationship connections. Ultimately, it is our responsibility to introduce this as a healthy part of the food chain. Discussion of what happens to the fox if he can't find any rabbits as well as to the rabbits if there aren't any foxes, helps children gain a sense of this delicate balance in nature.

---

- What can the fox do to make sure he gets some dinner?
- What happens to foxes who don't catch any rabbits?
- What would happen if there weren't any foxes? Rabbits?

## ADDITIONAL ACTIVITIES

- **GROUP**—Encourage children to talk about other predator-and-prey relationships. Be sure to stress the need for both predator and prey in the food chain. Discuss the proportional numbers of predator and prey types in the wild. Many prey animals must be present for few predator animals to survive.
- **GROUP**—Demonstrate how each link of the food chain is dependent on the others by making a paper chain with pictures of the food chain on it. For example, the first chain could be a corn plant, followed by a grasshopper, frog, and, finally, a person on the last link. Remove the link in the middle and talk about what happens to the others in the food chain.
- **MANIPULATIVE**—Play a matching game where children match pictures of predators with their prey. Place the game in the manipulative area during self-selected activity time.
- **NUTRITION**—Discuss the source of various food items eaten during lunch or snack.
- **STORY**—Read *Eat and Be Eaten* by Iela Mari (see appendix A on p. 181). This picture book features animals chasing smaller animals.

# Harvest Time ❧ Nutrition

## DID YOU KNOW?

Wheat is a form of domestic grass, and is an important food for animals and people. The wheat harvest generally occurs in early to mid summer. This activity will help children identify food products made from wheat and enable them to see and taste it in its raw form.

## MATERIALS

- **SEVERAL WHEAT PLANTS (ENTIRE PLANT, INCLUDING ROOTS)**
- **2 CUPS FLOUR**
- **4 TEASPOONS BAKING POWDER**
- **½ TEASPOON SALT**
- **½ TEASPOON CREAM OF TARTAR**
- **2 TEASPOONS SUGAR**
- **½ CUP BUTTER OR MARGARINE**
- **⅔ CUP MILK**
- **LARGE BOWL**
- **MEASURING CUPS AND SPOONS**
- **COOKIE SHEET**
- **BISCUIT CUTTER**
- **LARGE SPOON FOR STIRRING**

## ACTIVITY

1. Place the wheat plant in the science area during self-selected activity time. Encourage children to examine the plant and discuss where it came from.

2. Encourage children to help make biscuits with the listed ingredients. Let them sift and mix the ingredients, knead the biscuits, cut them out, and place them on the cookie sheet.

3. As children work, encourage them to taste the raw wheat and flour. Discuss the difference in taste, texture, color, and other characteristics.

4. Bake biscuits on an ungreased cookie sheet for 10 to 12 minutes at 450°F. The recipe makes about a dozen medium-sized biscuits.

5. Serve biscuits for snack. Talk about the process of change the wheat underwent, from wheat plant to biscuit. Ask children open-ended questions like these:
   - How do you think wheat is made into flour?

- What else do farmers use wheat for?
- What are some other ways people use flour?

## ADDITIONAL ACTIVITIES

- **BLOCK**—Provide toy tractors and various pieces of farm equipment in the play area during self-selected activity time.

- **BULLETIN BOARD**—Display pictures of the wheat in its various stages—from seed to plant to flour to products.

- **FIELD TRIP**—Visit a farm where wheat is being harvested.

- **MANIPULATIVE**—Provide a bowl and pestle for children to try grinding the wheat into flour. Experiment with other grinding utensils, such as rocks and large sticks.

- **NUTRITION**—Use cookie cutters shaped like farm animals to cut out biscuits. Talk about why the farmer has each animal and how each animal helps people.

- **NUTRITION**—Provide honey with the biscuits and discuss where honey comes from.

- **PRETEND PLAY**—Place farmer dress-up clothes (such as overalls, caps, and bandannas) in the dress-up area during self-selected activity time.

- **STORY**—Select one of the many versions of *The Little Red Hen* to read and act out.

# Rock Collection ▣ Field Trip

## DID YOU KNOW?

The earth's crust is made up of rocks and rock material. We dig it, tunnel through it, and build on it. All plants, animals, and people live on it. The earth's crust forms three-tenths of the entire surface of the earth; the rest is covered by water. This experience will enable children to begin their exploration of rocks.

## MATERIALS

- **EGG CARTONS (ONE PER CHILD)**
- **COTTON BATTING**
- **OUTDOOR AREA WITH INTERESTING ROCKS**
- **MAGNIFYING GLASSES**
- **SCIENCE JOURNALS**
- **COLORED PENCILS**
- **CAMERA (OPTIONAL)**

## ACTIVITY

1. Before passing out the egg cartons, label each one with the location where the rocks will be collected. This will help children remember and differentiate between the rocks collected in different areas.

2. Provide each child with an egg carton and cotton batting. Suggest that each section be lined with cotton batting to protect the rock specimens. This will provide children with space to collect and display at least twelve rocks.

3. Talk with children about examining their rocks and making wise choices. They should be responsible for carrying and keeping track of their own collection. Ask children open-ended questions like these:
   - What are some of the colors you have found in the rocks you collected?
   - Do you have any rocks that look like your friends' rocks?
   - Where do you think rocks come from?
   - What do you plan to do with your collection?

4. As children take a break or finish their collections, encourage them to examine their rocks with the magnifying glasses and compare their discoveries with those of other children.

5. Science journals should be used for documentation of data and sketches. Model collecting,

examining, writing, and sketching for the children, but accept anything they identify as writing or drawing.

6. Photographs can be used to identify and help remember key features of the collection site, and will provide a basis of comparing other sites visited later.

## ADDITIONAL ACTIVITIES

- **ART**—During self-selected activity time, encourage children to create granite paper using glitter crayons. Use magnifying glasses to examine the various colors in several different pieces of granite. Encourage children to select three or four glitter crayons that match the colors in the granite rocks. Sit on a rough concrete surface and make a rubbing of the concrete beginning with the lightest color. Cover the paper completely. Move the paper slightly as each color is added from lightest to darkest. Try placing the granite on top of the paper when finished and see how well the rock matches the paper.
- **FIELD TRIP**—Visit an art or anthropology museum and explore how people use rocks in art and tool making.
- **MANIPULATIVE**—During self-selected activity time, choose from one of the following:
  - Encourage children to match rocks by size, color, shape, or texture.
- Provide semiprecious gemstones purchased in a rock or nature shop for children to touch, explore, and sort. Children can match them with pictures on a poster or in a book.
  - Collect double pictures of various types of rocks and minerals to make a memory game.
- **OUTSIDE**—Encourage children to explore the play yard after a hard rain to see if any new rocks have been exposed.
- **PRETEND PLAY**—Place boots, goggles, maps, small rock hammers, notebooks, pencils, gloves, compasses, backpacks, rock and mineral field guides, and canteens or water bottles in the play area for children to pretend to be geologists during self-selected activity time.
- **SCIENCE**—During self-selected activity time, choose from one of the following:
  - Add a balance scale to the science area for children to weigh and compare their specimens.
  - Place a "feely box" in the science area with rocks of several different textures (smooth, rough, sharp, porous, and so on) for children to identify texture through touch alone.
- **STORY**—Read *Everybody Needs a Rock* by Byrd Baylor (see appendix A on p. 185). This story describes the qualities to look for in selecting the perfect rock.

# What Is a Mineral? ▣ Science

## DID YOU KNOW?

Minerals are naturally occurring substances that are neither animal nor plant, and they are made of many particles. Rocks are made up of one or more different minerals. Gold, silver, and copper are all considered minerals. This activity will expose children to minerals—how they are mined and how people use them.

## MATERIALS

- **GOLD CHAIN**
- **COPPER PENNY**
- **COPPER PIPE**
- **SILVER COIN**
- **SPRAY PAINT (1 CAN EACH OF GOLD, SILVER, AND COPPER)**
- **GRAVEL**
- **SANDBOX**
- **SIFTERS**
- **SHOVELS**
- **MAGNIFYING GLASSES**
- **SCIENCE JOURNALS**

## ACTIVITY

1. Paint gravel with gold, silver, and copper spray paint. Bury the gold, silver, and copper nuggets throughout the sandbox.

2. Provide sifters and shovels for children to mine for minerals during self-selected activity time. Compare this to how minerals are obtained.

3. Inside the classroom, arrange an attractive display of the gold, silver, and copper items in the science area. If available, add actual samples of gold, copper, and silver. Be sure to place the arrangement where children can handle and examine the specimens.

4. During self-selected activity time, encourage children to examine the items and talk about them. Ask children open-ended questions like these:
   - How did this mineral become this shape?
   - How do people use this mineral?
   - What would happen if people used up all of the minerals?

   Provide science journals for children to document their observations. Discuss how the minerals mined on the playground change into items like those in the science area.

## Ideas from Sherri's Classroom

This is a popular activity with the children—it allows them to experience one way minerals are extracted from the earth. It also generally presents an opportunity to explore the three levels of conservation. Of course, the children always mine the sandbox until there are no more minerals—or at least they can't find any more. Some children always end up with more nuggets than others. Children get upset and a discussion about fair distribution of the resource results. This is a conservation concept and the three levels of conservation—preservation, restoration, and management—always enter the discussion. Some children want to put all of the nuggets back and find them all over again, and others want to create more nuggets. Some children think no one should keep the nuggets found—they should all go into a basket in the classroom. Other children think each person should get to keep whatever they find. There isn't a right answer, but it is always a debate about how to distribute the nuggets.

## Additional Activities

- **Group**—Label objects around the room made from minerals.
- **Pretend play**—During self-selected activity time put out mining clothes for children to wear and pretend to mine for minerals. Clothing might include coveralls and hardhats or bike helmets with flashlights fastened to the top with duct tape.
- **Science**—During self-selected activity time, have children create a crystal garden by mixing ¼ cup salt, ¼ cup bluing, and ¼ cup ammonia in a jar and pour over six charcoal briquets placed in a pie plate. Repeat the solution every two days. Provide science journals and magnifying glasses for children to document their observations. This activity should be done in a well-ventilated area or outside to prevent inhalation of fumes.

# Concrete Hand Impressions ▨ Outside

## DID YOU KNOW?

Concrete is made up of water and different kinds of rocks and minerals. This activity will show children one way people use rocks and minerals.

## MATERIALS

- **HEAVY PAPER PLATES (LARGE ENOUGH FOR A CHILD'S HAND)**
- **SAND**
- **CEMENT**
- **WATER**
- **PAPER TOWELS**
- **BUCKET**

## ACTIVITY

1. During self-selected activity time, encourage children to fill a paper plate with wet sand.

2. After filling the plate with sand, children should press one hand into the sand, making an impression. If sand is too dry to retain the impression, mix water with sand and try again.

3. In a bucket, mix four parts sand and one part cement. Add water until the mixture is the consistency of thick cake batter.

4. Encourage children to fill their impressions with concrete.

5. After the concrete dries, children can paint or decorate their concrete impressions.

6. Lead a discussion about concrete by asking children open-ended questions like these:
   - How many ways can you think of that people use concrete?
   - Which minerals do you think it takes to make concrete?
   - What is the purpose of a concrete truck?
   - How does concrete change the land?

## ADDITIONAL ACTIVITIES

- **BLOCK**—During self-selected activity time, provide toy concrete trucks in the block area for children to pretend to mix and pour concrete. Provide large pieces of gravel for children to try building various stone structures.
- **FIELD TRIP**—Visit a rock quarry.
- **FIELD TRIP**—Visit a site where workers are pouring concrete. As you are traveling to the site, look for rocks and stone used in your town's architecture.
- **PRETEND PLAY**—During self-selected activity time, put out coveralls in the dress-up area for children to pretend to be concrete truck drivers hauling concrete.
- **PRETEND PLAY**—Place pea gravel in a sensory table or large tub for children to pour, dump, feel, and experiment with during self-selected activity time. Compare this with the gravel in the parking lot or driveway.

- **SCIENCE**—Set up a rock display, using a special table covering, unbreakable mirrors, baskets for displaying rocks, and so on. Be sure to include several magnifying glasses, flashlights, and balance scales so children can explore and investigate the rocks during self-selected activity time. Provide science journals and colored pencils in the area for children to write about and sketch rocks.
- **STORY**—Read *Building a House* by Byron Barton (see appendix A on p. 184). This book demonstrates the steps followed in building a house.

# The Quest for Shiny Rocks ▤ Science

## DID YOU KNOW?

Rocks and pebbles roll along river and stream beds and become smooth and round after many, many years. However, to appreciate their full beauty, they need to be polished or varnished.

Children will enjoy experimenting with various solutions to make their rock collections shiny.

## MATERIALS

- **ROCKS THAT ARE SMOOTH, ROUND, AND BEAUTIFUL (COLLECT THESE ALONG AND IN STREAMS OR RIVER BEDS)**
- **TOOTHPASTE (GEL AND PASTE)**
- **BABY OIL**
- **PETROLEUM JELLY**
- **PASTE WAX**
- **BAKING SODA**
- **VINEGAR**
- **OLD TOOTHBRUSHES**
- **TWO TUBS OF WATER**
- **PAPER TOWELS**
- **AUDIO RECORDER AND BLANK CASSETTE (OPTIONAL)**
- **CAMERA (OPTIONAL)**

## ACTIVITY

1. During group time, discuss with children how they think rocks become shiny. Using the audio recorder, record their conversations. Challenge children to think of materials they think might make rocks shiny.

2. Place the rocks, water, paper towels, and toothbrushes in the science area. On alternating days, provide a different solution for shining the rocks. One day the children might try gel toothpaste, the next day baby oil, and so on. Encourage children to revisit this area during self-selected activity time.

3. Photograph the rocks each day after children have finished shining them.

4. As children try the various polishes, challenge them to think of other ideas for polishing the rocks. Provide other materials as they are suggested. Ask children open-ended questions like these:
   - What happens when you use this type of polish on the rocks?
   - What happens if you rub harder?
   - Which shining solution do you think works the best?

## Ideas from Sherri's Classroom

In a previous summer, one group of children was particularly intrigued with how rocks become shiny. It takes about a month to tumble rocks in a rock tumbler so, although we had started the tumbler, they were still seeking answers. They asked everyone they knew until finally someone told them that she heard that toothpaste cleaned rocks. They tried both gel and paste but this small suggestion started them thinking. The toothpaste cleaned the rocks but didn't make them shiny. They knew about the chemical reaction of baking soda and vinegar from past experiments. They thought this might have some effect on the rocks; however, they still weren't shiny enough. From here, they tried petroleum jelly and baby oil. Although they made the rocks shiny, they weren't pleasant to touch. Wax was their next suggestion and provided the most satisfactory results for this particular group of children. That summer, they spent a great deal of time polishing rocks. It was a soothing, successful activity that they requested again and again. Your children may have different ideas and experiences about how to make rocks shiny. Be sure to support their theory testing by gathering any materials they need.

---

- How do you think jewelry stores keep gemstones shiny?
- Where could we find out how to make rocks shiny?

5. Transcribe the audio recordings from each of your discussions to help decide further areas of exploration and study.

## Additional Activities

- **Art**—During self-selected activity time, provide spray bottles of water for children to keep rocks wet (shiny). Set out watercolors, brushes, and paper for children to paint rock portraits.
- **Bulletin board**—Display the pictures and transcriptions of children's discussions to encourage revisiting and reflecting on the experience.
- **Science**—Encourage children to help prepare and set up a rock tumbler (a machine with a rotating drum that polishes rocks). Pebbles or rocks are placed inside the tumbler, along with grinding powder and water. The process is noisy and should be done in an isolated spot, and it takes several weeks to successfully polish the pebbles and rocks. Once the process is completed, the tumbled rocks can be used for sorting, jewelry making, or just admiring.

# APPENDIX A

## Children's Literature

This appendix lists fiction and nonfiction children's literature appropriate for classroom use, arranged by theme or topic. Criteria for literature selection included realistic portrayal of animals in their natural habitat, accurate and factual information, relevance to concepts children might construct from this topic, and overall quality. Literature depicting animals with human characteristics was generally not included. One of the developmental challenges that preschool children struggle with is the distinction between reality and fantasy. The use of realistic literature for conservation discussions encourages children to develop attitudes and ideas about nature that are not tied to emotions. This is not to say that fiction portraying animals with human characteristics is not appropriate but rather to encourage those sharing that literature to discuss the pretend versus real nature of all stories shared with young children.

### Air and Water

Ariane. 1996. *Small cloud*. New York: Walker and Company.

Asch, Frank. 1997. *Water*. New York: Scholastic, Inc.

Bacon, Ron. 1993. *Wind*. New York: Scholastic, Inc.

Branley, Franlyn M. 1990. *Air is all around you*. New York: HarperCollins.

Canizares, Susan, and Pamela Chanko. 1998. *Water*. New York: Scholastic, Inc.

Cole, Joanna. 1988. *The Magic School Bus: At the waterworks*. New York: Scholastic, Inc.

Dorrors, Arthur. 1993. *Follow the water from brook to ocean*. New York: HarperCollins.

Ets, Marie H. 1978. *Gilberto and the wind*. New York: Viking Press.

Gibbons, Gail. 1989. *Catch the wind! All about kites*. Boston: Little, Brown and Company.

Hathorn, Libby. 1996. *The wonder thing*. New York: Houghton Mifflin.

Hutchins, Pat. 1974. *The wind blew*. New York: Scholastic, Inc.

Jeunesse, Gallimard, and Pierre-Marie Valat. 1996. *Water: A First Discovery Book*. New York: Scholastic, Inc.

McKinney, Barbara S. 1998. *A drop around the world*. Nevada City, CA: Dawn Publishers.

Mizumura, Kazue. 1966. *I see the winds*. New York: Thomas Y. Crowell Company.

Pluckrose, Henry. 1994. *Walkabout in the air*. Chicago: Children's Press.

Ruis, Maria, and J. M. Parramon. 1985. *The four elements: Air*. Woodbury, NY: Barron's.

Vendrell, Carme Sole, and J. M. Parramon. 1985. *The four elements: Water*. Woodbury, NY: Barron's.

Wick, Walter. 1997. *A drop of water*. New York: Scholastic, Inc.

### Amphibians and Reptiles

Arnosky, Jim. 1994. *All about alligators*. New York: Scholastic, Inc.

———. 1997. *All about rattlesnakes*. New York: Scholastic, Inc.

———. 2000. *All about turtles*. New York: Scholastic, Inc.

———. 2002. *All about frogs*. New York: Scholastic, Inc.

Behler, John. 1999. *National Audubon Society: First Field Guide: Reptiles*. New York: Scholastic, Inc.

Berger, Melvin, and Gilda Berger. 2003. *Frogs live on logs*. New York: Scholastic, Inc.

———. 2003. *Snakes live in grass*. New York: Scholastic, Inc.

Berkes, Marianne. 2002. *Marsh music*. Brookfield, CT: Millbrook Press.

Bernhard, Emery. 1995. *Salamanders*. New York: Holiday House.

Brown, Ruth. 1999. *Toad*. New York: Penguin Putnam.

Clarke, Barry. 2000. *Amphibian: Eyewitness Books*. New York: DK Publishing.

Cole, Joanna. 1981. *A snake's body*. New York: Trumpet Club.

Craig, Janet. 1989. *Now I know about turtles*. Mahwah, NJ: Troll Associates.

Demuth, Patricia. 1993. *Snakes*. New York: Putnam Publishing.

Dewey, Jennifer. 1989. *Can you find me? A book about animal camouflage*. New York: Scholastic, Inc.

Dussling, Jennifer. 1998. *Eyewitness Readers: Slinky, scaly snakes!* New York: DK Publishing.

Fichter, George S. 1993. *A Golden Junior Guide: Snakes and lizards*. New York: A Golden Book.

Flack, Marjorie. 1959. *Tim tadpole and the great bullfrog*. Garden City, NY: Doubleday.

Freschet, Berniece. 1971. *Turtle pond*. New York: Charles Scribner's Sons.

George, William T. 1989. *Box turtle at Long Pond*. New York: Greenwillow Books.

Gibbons, Gail. 1994. *Frogs*. New York: Holiday House.

Gross, Ruth Belov. 1991. *Snakes*. New York: Simon and Schuster.

Hawes, Judy. 1996. *Why frogs are wet*. New York: HarperCollins.

Heller, Ruth. 1995. *How to hide a meadow frog and other amphibians*. New York: Grosset and Dunlap.

———. 1999. *Chickens aren't the only ones*. New York: Penguin Putnam.

Hoffman, Mary. 1987. *Animals in the wild: Snake*. New York: Scholastic, Inc.

Jeunesse, Gallimard, and Daniel Moignot. 1997. *Frogs: A First Discovery Book*. New York: Scholastic, Inc.

Jeunesse, Gallimard, and Gilbert Houbre. 1998. *Turtles and snails: A First Discovery Book*. New York: Scholastic, Inc.

Julivert, Maria Angels. 1993. *The fascinating world of snakes*. Hauppauge, NY: Barron's.

Kalan, Robert. 1989. *Jump, frog, jump!* New York: Scholastic, Inc.

Kalman, Bobby, and Tammy Everts. 1994. *Frogs and toads*. New York: Crabtree Publishing.

Lasky, Kathryn. 1997. *Pond year*. Cambridge, MA: Candlewick Press.

Lauber, Patricia. 1989. *Snakes are hunters*. New York: HarperCollins.

Maestro, Betsy. 1997. *Take a look at snakes*. New York: Scholastic, Inc.

Markle, Sandra. 1995. *Outside and inside snakes*. New York: Atheneum.

———. 1998. *Outside and inside alligators*. New York: Scholastic, Inc.

Matero, Robert. 1997. *Eyes on nature: Lizards*. Chicago: Kidsbooks.

Mazer, Anne. 1991. *The salamander room*. New York: Knopf.

O'Neill, Amanda. 1996. *I wonder why snakes shed their skin and other questions about reptiles*. New York: Larousse Kingfisher Chambers.

Pallotta, Jerry. 1990. *The frog alphabet book*. Watertown, MA: Charlesbridge.

Parker, Nancy Winslow, and Joan Richards Wright. 1990. *Frogs, toads, lizards, and salamanders*. New York: Greenwillow Books.

Parker, Steve. 1999. *It's a frog's life*. Pleasantville, NY: Reader's Digest Children's Books.

Parsons, Alexandra. 1990. *Eyewitness Juniors: Amazing snakes*. New York: Random House.

Petty, Kate. 1998. *I didn't know that crocodiles yawn to keep cool*. Brookfield, CT: Copper Beech Books.

Pfeffer, Wendy. 1994. *From tadpole to frog*. New York: HarperCollins.

Porte, Barbara Ann. 1999. *Tale of a tadpole*. New York: Scholastic, Inc.

Smith, Trevor. 1991. *Eyewitness Juniors: Amazing lizards*. New York: Knopf.

Taylor, Barbara. 1999. *Nature Watch: Snakes*. London: Anness.

Royston, Angela. 1991. *What's inside? Small animals*. New York: DK Publishing.

Wallace, Karen. 1998. *Tale of a tadpole: Eyewitness Readers*. New York: DK Publishing.

Winer, Yvonne. 2003. *Frogs sing songs*. Watertown, MA: Charlesbridge.

## Animals in General

Arnosky, Jim. 1978. *Outdoor on foot*. New York: Coward, McCann and Geogehan, Inc.

———. 1995. *I see animals hiding*. New York: Scholastic, Inc.

———. 2002. *Field trips: Bug hunting, animal tracking, shore walking, bird-watching with Jim Arnosky*. New York: HarperCollins.

Barchas, Sarah E. 1993. *I was walking down the road*. New York: Scholastic, Inc.

Barrett, Judi. 1989. *Animals should definitely not act like people*. New York: Simon and Schuster.

———. 1989. *Animals should definitely not wear clothing*. New York: Simon and Schuster.

Brown, Margaret Wise. 1959. *Nibble nibble: Poems for children*. New York: HarperCollins.

Canizares, Susan, and Pamela Chanko. 1998. *Who's hiding?* New York: Scholastic, Inc.

Dewey, Jennifer. 1989. *Can you find me? A book about animal camouflage*. New York: Scholastic, Inc.

Dorros, Arthur. 1991. *Animal tracks*. New York: Scholastic, Inc.

George, Linsday Barrett. 1995. *In the woods: Who's been here?* New York: William Morrow.

Kitchen, Bert. 1994. *Somewhere today*. Cambridge, MA: Candlewick Press.

Machotka, Hana. 1991. *What neat feet!* New York: Scholastic, Inc.

Nail, James. 1996. *Whose tracks are these? A clue book of familiar forest animals*. New York: Roberts Rinehart.

Roop, Peter, and Connie Roop. 1992. *One earth, a multitude of creatures*. New York: Walker and Co.

Royston, Angela. 1992. *What's inside? Small animals*. New York: Dorling Kindersley

Ryan, Pam Munoz. 1999. *A pinky is a baby mouse and other baby animal names*. New York: Scholastic, Inc.

Ryder, Joanna. 1988. *The snail's spell*. New York: Viking Press.

————. 2001. *A fawn in the grass*. New York: Henry Holt.

Schwartz, Elizabeth. 1964. *When animals are babies*. New York: Holiday House.

Selsam, Millicent Ellis. 1995. *How to be a nature detective*. New York: HarperCollins.

Singer, Marilyn. 1989. *Turtle in July*. New York: Macmillan.

Yee, Wong Herbert. 2003. *Tracks in the snow*. New York: Henry Holt.

## Aquatic Life

Arnosky, Jim. 1997. *Watching water birds*. Washington, DC: National Geographic Society.

————. 1999. *Otters under water*. New York: Penguin Putnam.

Berger, Melvin. 1994. *Oil spill!* New York: HarperCollins.

Cooper, Ann. 1998. *Around the pond*. Boulder, CO: Robert Rinehart.

Cristini, Ermanno, and Luigi Puricelli. 1989. *In the pond*. New York: Simon and Schuster.

Fleming, Denise. 1993. *In the small, small pond*. New York: Henry Holt.

Halpern, Shari. 1995. *My river*. New York: Scholastic, Inc.

Jeunesse, Gallimard, and Laura Bour. 1993. *The river: A First Discovery Book*. New York: Scholastic, Inc.

Lasky, Kathryn. 1997. *Pond year*. Cambridge, MA: Candlewick Press.

Mazer, Anne. 1991. *The salamander room*. New York: Knopf.

Parker, Nancy Winslow, and Joan Richards Wright. 1990. *Frogs, toads, lizards, and salamanders*. New York: Greenwillow Books.

Rosen, Michael J. 1994. *All eyes on the pond*. New York: Hyperion Books.

Taylor, Barbara. 1998. *Look closer: Pond life*. New York: DK Publishing.

Tresselt, Alvin. 1970. *The beaver pond*. New York: Lothrop, Lee and Shepard.

## Birds

Arnosky, Jim. 1997. *Watching water birds*. Washington, DC: National Geographic Society.

————. 1998. *All about turkeys*. New York: Scholastic, Inc.

————. 1999. *All about owls*. New York: Scholastic, Inc.

————. 1999. *All night near the water*. New York: Paper Star.

Back, Christine, and Jens Olesen. 1984. *Chicken and egg*. New York: Silver Burdett.

Berger, Melvin, and Gilda Berger. 2003. *Owls live in trees*. New York: Scholastic, Inc.

Bishop, Nic. 1997. *The secrets of animal flight*. New York: Houghton Mifflin.

Brown, Margaret Wise. 1995. *The dead bird*. New York: HarperCollins.

Browne, Vee. 1995. *Animal lore and legend: Owl*. New York: Scholastic, Inc.

Bunting, Eve. 1994. *Night tree*. San Diego: Harcourt Brace.

————. 1996. *Secret place*. New York: Clarion Books.

Carlstrom, Nancy White. 1991. *Goodbye geese*. New York: Scholastic, Inc.

Cherry, Lynne. 1997. *Flute's journey: The life of a wood thrush*. New York: Harcourt Brace.

Collard, Sneed B., III. 2002. *Beaks!* Watertown, MA: Charlesbridge.

Curran, Eileen. 1985. *Bird's nests*. Mahwah, NJ: Troll Associates.

Delafosse, Claude, and Rene Mettler. 1990. *Birds: A First Discovery Book*. New York: Scholastic, Inc.

Downing, Julie. 1991. *White snow, blue feather*. New York: Simon and Schuster.

Ehlert, Lois. 1990. *Feathers for lunch*. Orlando, FL: Harcourt Brace.

Florian, Douglas. 1996. *On the wing*. New York: Harcourt Brace.

Galinsky, Ellen. 1977. *The baby cardinal*. New York: G. P. Putnam's Sons.

Gans, Roma. 1996. *How do birds find their way?* New York: HarperCollins.

Garelick, May. 1970. *What's inside?* New York: William R. Scott.

Gibbons, Gail. 2001. *Ducks!* New York: Holiday House.

Heller, Ruth. 1999. *Chicken's aren't the only ones*. New York: Penguin Putnam.

Jenkins, Priscilla Betz. 1995. *A nest full of eggs*. New York: HarperCollins.

Jeunesse, Gallimard, and Pascale de Bourgoing. 1992. *The egg: A First Discovery Book*. New York: Scholastic, Inc.

Jeunesse, Gallimard, and Sylvaine Perols. 1998. *Night creatures: A First Discovery Book*. New York: Scholastic, Inc.

Johnston, Tony. 2002. *The barn owls*. New York: Charlesbridge.

Jonas, Ann. 1999. *Bird talk*. New York: Greenwillow Books.

Kuchalla, Susan. 1982. *Now I know birds*. Mahwah, NJ: Troll Associates.

Mainwaring, Jane. 1989. *My feather*. New York: Doubleday.

Mazzola, Frank Jr. 1997. *Counting is for the birds*. New York: Charlesbridge.

Oppenheim, Joanne. 1990. *Have you seen birds?* New York: Scholastic, Inc.

Ryder, Joanne. 1989. *Catching the wind*. New York: William Morrow.

Sill, Cathryn. 1997. *About birds: A guide for children*. Atlanta: Peachtree Publishers.

Tafuri, Nancy. 1984. *Have you seen my duckling?* New York: Greenwillow Books.

Weidensaul, Scott. 1998. *National Audubon Society: First Field Guide: Birds*. New York: Scholastic, Inc.

Wildsmith, Brian. 1981. *Birds*. New York: Oxford University Press.

Yolen, Jane. 1987. *Owl moon*. New York: Scholastic, Inc.

## Camping

Baylor, Byrd. 1978. *The other way to listen*. New York: Simon and Schuster.

Bunting, Eve. 1996. *I don't want to go to camp*. Honesdale, PA: Boyds Mills Press.

Brillhart, Julie. 1997. *When daddy took us camping*. Morton Grove, IL: Albert Whitman and Company.

Carrick, Carol, and Donald Carrick. 1973. *Sleep out*. New York: Clarion Books.

George, Kristine O'Connell. 2001. *Toasting marshmallows: Camping poems*. New York: Clarion Books.

Hafner, Marylin. 2001. *Molly and Emmett's camping adventure*. Columbus, OH: Carus Publishing.

Huneck, Stephen. 2001. *Sally goes to the mountains*. New York: Harry N. Abrams, Inc.

Partridge, Elizabeth. 2003. *Whistling*. New York: Greenwillow Books.

Ruurs, Margriet. 2001. *When we go camping*. Plattsburgh, NY: Tundra Books.

Singer, Marilyn. 2002. *Quiet night*. New York: Clarion Books.

Smith, Tim. 1997. *Buck Wilder's small twig hiking and camping guide: A complete introduction to the world of hiking and camping for small twigs of all ages*. Williamsburg, MI: Alexander & Smith Publishing.

Stern, Maggie. 1998. *Acorn magic*. New York: Greenwillow Books.

Van Dusen, Chris. 2003. *A camping spree with Mr. Magee*. San Francisco: Chronicle Books.

Williams, Vera B. 1981. *Three days on a river in a red canoe*. New York: Mulberry Books.

Wolff, Ashley. 1999. *Stella and Roy go camping*. New York: Dutton Children's Books.

## Caves

Angeletti, Roberta. 1999. *A journey through time: The cave painter of Lascaux*. New York: Oxford University Press.

Butler, Daphne. 1991. *First look: Under the ground*. Milwaukee: Gareth Stevens.

Davies, Nicola. 2001. *Bat loves the night*. Cambridge, MA: Candlewick Press.

Delafosse, Claude, and Gallimard Jeunesse. 1998. *Cave: Hidden world*. New York: Scholastic, Inc.

Gans, Roma, and Biulio Maestro. 1990. *Caves*. New York: HarperCollins.

Gibbons, Gail. 1996. *Caves and Caverns*. New York: Harcourt Brace

———. 1999. *Bats*. New York: Scholastic, Inc.

Greenberg, Judith E., and Helen Carey. 1990. *Caves*. Milwaukee: Raintree Publishers.

Gunzi, Christine. 2000. *Look closer: Cave life*. New York: DK Publishing.

Jeunesse, Gallimard, and Sylvaine Perols. 1998. *Night creatures: A First Discovery Book*. New York: Scholastic, Inc.

Kerbo, Ronald C. 1981. *Caves*. Chicago: Childrens Press.

Schultz, Ronald. 1993. *Looking inside caves and caverns*. Santa Fe, NM: John Muir Publications.

Siebert, Diane. 2000. *Caves*. New York: HarperCollins.

Silver, Donald M. 1997. *One small square: Cave*. New York: McGraw-Hill.

Zim, Herbert Spencer. 1978. *Caves and life*. New York: William Morrow.

## Energy

Berger, Melvin. 1990. *Switch on, switch off*. New York: HarperCollins.

———. 1995. *All about electricity*. New York: Scholastic, Inc.

Berger, Samantha, and Pamela Chanko. 1999. *Electricity*. New York: Scholastic, Inc.

Bradley, Kimberly Brubaker. 2003. *Energy makes things happen.* New York: HarperCollins.

Cole, Joanna, and Bruce Degen. 1997. *The Magic School Bus and the electric field trip.* New York: Scholastic, Inc.

DeRegniers, Beatrice Schenk. 1961. *Who likes the sun?* New York: Harcourt Brace.

Gibbons, Gail. 1983. *Sun up, sun down.* New York: Harcourt Brace.

Lionni, Leo. 1987. *Alexander and the wind-up mouse.* New York: Knopf.

Pondendorf, Illa. 1963. *The true book of energy.* Chicago: Children's Press.

## Fish

Berger, Melvin, and Gilda Berger. 2003. *Fish live in water.* New York: Scholastic, Inc.

Cook, Bernadine. 1956. *The little fish that got away.* New York: David McKay Co.

Dewey, Jennifer. 1989. *Can you find me? A book about animal camouflage.* New York: Scholastic, Inc.

Heller, Ruth. 1999. *Chickens aren't the only ones.* New York: Penguin Putnam.

Jeunesse, Gallimard, Claude Delafosse, and Sabine Krawczyk. 1998. *Fish: A First Discovery Book.* New York: Scholastic, Inc.

Kalan, Robert. 1992. *Blue sea.* New York: William Morrow.

Lasky, Kathryn. 1997. *Pond year.* Cambridge, MA: Candlewick Press.

Lionni, Leo. 1992. *Swimmy.* New York: Knopf.

Pallotta, Jerry. 1996. *The freshwater alphabet book.* New York: Scholastic, Inc.

Pfeffer, Wendy. 1996. *What's it like to be a fish?* New York: HarperCollins.

Smith, Tim. 1995. *Buck Wilder's small fry fishing guide: A complete introduction to the world of fishing for small fry of all ages.* Williamsburg, MI: Alexander & Smith Publishing.

Wildsmith, Brian. 1987. *Fishes.* New York: Oxford University Press.

## Food Chains

Dewey, Jennifer. 1995. *Can you find me? A book about animal camouflage.* New York: Scholastic, Inc.

Facklam, Margery. 1999. *Bugs for lunch.* New York: Scholastic.

Freschet, Berniece. 1968. *The old bullfrog.* New York: Charles Scribner's Sons.

Godkin, Celia. 1998. *What about ladybugs?* San Francisco: Sierra Club.

Hutchings, Pat. 1972. *Rosie's walk.* New York: Simon and Schuster.

Jenkins, Steve. 1997. *What do you do when something wants to eat you?* New York: Houghton Mifflin.

Kalan, Robert. 1991. *Jump, frog, jump!* New York: William Morrow.

Lauber, Patricia. 1995. *Who eats what? Food chains and food webs.* New York: HarperCollins.

Mari, Iela. 1980. *Eat and be eaten.* Woodbury, NY: Barron's.

Noll, Sally. 1990. *Watch where you go.* New York: Penguin Putnam.

Schoenherr, John. 1968. *The barn.* New York: Little, Brown and Company.

## Habitat

Albert, Richard E. 1994. *Alejandro's gift.* San Francisco: Chronicle Books.

Arnosky, Jim. 1997. *Crinkleroot's guide to knowing animal habitats.* New York: Simon and Schuster.

Ashman, Linda. 2001. *Castles, caves, and honeycombs.* San Diego: Harcourt, Inc.

Canizares, Susan, and Mary Reid. 1998. *Homes in the ground.* New York: Scholastic, Inc.

Cristini, Ermanno, and Luigi Puricelli. 1981. *In my garden.* New York: Scholastic, Inc.

———. 1983. *In the woods.* New York: Scholastic, Inc.

———. 1989. *In the pond.* New York: Simon and Schuster.

Dunphy, Madeleine. 1996. *Here is the wetland.* New York: Hyperion.

Dunrea, Oliver. 1989. *Deep down underground.* New York: Macmillan.

Fleming, Denise. 1993. *In the small, small pond.* New York: Henry Holt.

———. 1995. *In the tall, tall grass.* New York: Henry Holt.

Fleming, Maria. 1997. *How to build a home.* New York: Scholastic, Inc.

Fraser, Mary Ann. 1999. *Where are the night animals?* New York: HarperCollins

George, Lindsay B. 1995. *In the woods: Who's been here?* New York: Greenwillow Books.

Halpern, Shari. 1995. *My river.* New York: Scholastic, Inc.

Jeunesse, Gallimard, and Laura Bour. 1993. *The river: A First Discovery Book.* New York: Scholastic, Inc.

Jeunesse, Gallimard, and Pascale de Bourgoing. 1995. *Under the ground: A First Discovery Book.* New York: Scholastic, Inc.

Kitchen, Bert. 1995. *And so they build.* Cambridge, MA: Candlewick Press.

Lauber, Patricia. 1996. *You're aboard spaceship Earth.* New York: HarperCollins.

Parson, Alexandra. 1993. *What's inside? Animal homes.* New York: DK Publishing.

Pondendorf, Illa. 1960. *The true book of animal homes.* Chicago: Children's Press.

Reid, Mary. 1998. *Homes in the ground.* New York: Scholastic, Inc.

Rosen, Michael J. 1994. *All eyes on the pond.* New York: Hyperion Books.

Ryder, Joanne. 1990. *Under your feet.* New York: Four Winds Press.

————. 2000. *Each living thing.* San Diego: Gulliver Books.

Schreiber, Anne. 1994. *Log hotel.* New York: Scholastic, Inc.

Silver, Donald M. 1993. *One small square: Seashore.* New York: McGraw-Hill.

————. 1997. *One small square: Swamp.* New York: McGraw-Hill.

————. 1994. *One very small square: Life on a limb.* New York: W. H. Freeman.

————. 1994. *One very small square: Prairie dog town.* New York: W. H. Freeman.

————. 1995 *One very small square: Busy beaver pond.* New York: W. H. Freeman.

Thornhill, Jan. 1997. *Before and after: A book of nature timescapes.* Washington, DC: National Geographic Society.

Wilsmith, Brian. 1980. *Animal homes.* New York: Oxford University Press.

Zoehfeld, Kathleen Weidner. 1994. *What lives in a shell?* New York: HarperCollins.

## Harvest

Arnosky, Jim. 1991. *Raccoons and ripe corn.* New York: William Morrow.

Beskow, Elsa. 1997. *Pelle's new suit.* Beltsville, MD: Gryphon House.

Brandenberg, Aliki. 1976. *Corn is maize: The gift of the Indians.* New York: HarperCollins.

Davol, Marguerite W. 1992. *The heart of the wood.* New York: Simon and Schuster.

de Bourgoing, Pascale. 1991. *Fruit: A First Discovery Book.* New York: Scholastic, Inc.

————. 1994. *Vegetables in the garden: A First Discovery Book.* New York: Scholastic, Inc.

dePaola, Tomie. 1988. *Charlie needs a cloak.* New York: Aladdin Books.

————. 1989. *The popcorn book.* New York: Holiday House.

————. 1991. *Pancakes for breakfast.* San Diego: Harcourt Brace.

Ehlert, Lois. 1993. *Eating the alphabet.* New York: Harcourt Brace.

————. 2000. *Market day.* New York: Scholastic, Inc.

Gibbons, Gail. 1987. *The milk makers.* New York: Simon and Schuster.

Hausherr, Rosmarie. 1994. *What food is this?* New York: Scholastic, Inc.

Hutchings, Amy. 1994. *Picking apples and pumpkins.* New York: Scholastic, Inc.

Maestro, Betsy. 1992. *How do apples grow?* New York: HarperCollins.

Ray, Mary Lyn. 1992. *Pumpkins.* New York: Harcourt Brace.

Robbins, Ken. 1992. *Make me a peanut butter sandwich.* New York: Scholastic, Inc.

Rockwell, Anne. 1989. *Apples and pumpkins.* New York: Scholastic, Inc.

Tresselt, Alvin. 1951. *Autumn harvest.* New York: William Morrow.

Williams, Sherley Anne. 1992. *Working cotton.* New York: Trumpet Club.

## Insects

Berger, Melvin. 1998. *Chirping crickets.* New York: HarperCollins.

————. 2003. *Spinning spiders.* New York: HarperCollins.

Berger, Melvin, and Gilda Berger. 2003. *Bees live in hives.* New York: Scholastic, Inc.

Bishop, Nic. 1997. *The secrets of animal flight.* New York: Houghton Mifflin.

Brenner, Barbara, and Bernice Chardiet. 1995. *Where's that insect?* New York: Scholastic, Inc.

Brinckloe, Julie. 1986. *Fireflies!* New York: Simon and Schuster.

Brown, Ruth. 1988. *Ladybug, ladybug.* New York: Penguin Books.

Canizares, Susan, and Mary Reid. 1998. *Where do insects live?* New York: Scholastic, Inc.

Canizares, Susan, and Pamela Chanko. 1998. *What do insects do?* New York: Scholastic, Inc.

Cassie, Brian, and Jerry Pallotta. 1995. *The butterfly alphabet book.* New York: Charlesbridge.

Chinery, Michael. 1991. *Butterfly.* Mahwah, NJ: Troll Associates.

Cole, Joanna. 1998. *The Magic School Bus: Inside a beehive.* New York: Scholastic, Inc.

Conklin, Gladys. 1958. *I like caterpillars.* New York: Holiday House.

Cutts, David. 1989. *Look . . . a butterfly.* Mahwah, NJ: Troll Associates.

Cyrus, Kurt. 2001. *Oddhopper opera: A bug's garden of verses.* New York: Harcourt Brace.

de Bourgoing, Pascale. 1991. *The ladybug and other insects: A First Discovery Book.* New York: Scholastic, Inc.

Delafosse, Claude. 1997. *Butterflies: A First Discovery Book.* New York: Scholastic, Inc.

Dewey, Jennifer. 1989. *Can you find me? A book about animal camouflage.* New York: Scholastic, Inc.

Dorros, Arthur. 1998. *Ant cities.* New York: HarperCollins.

Ehlert, Lois. 2001. *Waiting for wings.* New York: Harcourt Brace.

Fisher, Aileen. 1986. *When it comes to bugs.* New York: Harper and Row.

Florian, Douglas. 1998. *Insectlopedia.* New York: Scholastic, Inc.

French, Vivian. 1993. *Caterpillar, caterpillar.* Cambridge, MA: Candlewick Press.

Frisky, Margaret. 1990. *Johnny and the monarch.* Chicago: Children's Press.

Fuhr, Uhr, and Raule Sautai. 1997. *Bees: A First Discovery Book.* New York: Scholastic, Inc.

Garelick, May. 1997. *Where does the butterfly go when it rains?* Greenvale, NY: Mondo Publishing.

Gibbons, Gail. 1989. *Monarch butterfly.* New York: Holiday House.

Godkin, Celia. 1998. *What about ladybugs?* San Francisco: Sierra Club.

Grassy, John. 1997. *Insects.* New York: Scholastic, Inc.

Hawes, Judy. 1991. *Fireflies in the night.* New York: HarperCollins.

———. 1967. *Ladybug, ladybug, fly away home.* New York: Thomas Y. Crowell.

Heiligman, Deborah. 1996. *From caterpillar to butterfly.* New York: HarperCollins.

Heller, Ruth. 1992. *How to hide a butterfly and other insects.* New York: Penguin Putnam.

———. 1999. *Chickens aren't the only ones.* New York: Penguin Putnam.

Hopkins, Lee Bennett. 1992. *Flit, flutter, fly! Poems about bugs and other crawly creatures.* New York: Delacorte Press.

Jenkins, Martin. 1996. *Wings, stings, and wriggly things.* Cambridge, MA: Candlewick Press.

Kilpatrick, Cathy. 1982. *Usborne First Nature: Creepy crawlies.* London: Usborne Publishing.

Lerner, Carol. 2002. *Butterflies in the garden.* New York: HarperCollins.

Ling, Mary. 1992. *See how they grow: Butterfly.* New York: DK Publishing.

Lionni, Leo. 1995. *Inch by inch.* New York: William Morrow.

Markle, Sandra. 1996. *Creepy crawly baby bugs.* New York: Scholastic, Inc.

McDonald, Megan. 1995. *Insects are my life.* New York: Orchard Books.

Oppenheim, Joanne. 1998. *Have you seen bugs?* New York: Scholastic, Inc.

Pallotta, Jerry. 1990. *The icky bug alphabet book.* Watertown, MA: Charlesbridge.

———. 1998. *The butterfly counting book.* New York: Scholastic, Inc.

Parker, Nancy Winslow, and Joan Richards Wright. 1988. *Bugs.* New York: William Morrow.

Reid, Mary, and Betsey Chessen. 1998. *Bugs, bugs, bugs!* New York: Scholastic, Inc.

Rockwell, Anne, and Harlow Rockwell. 1966. *Sally's caterpillar.* New York: *Parents'* Magazine Press.

Romanova, Natalia. 1992. *Once there was a tree.* New York: Penguin USA.

Royston, Angela. 1992. *What's inside? Insects.* New York: DK Publishing.

Ryder, Joanne. 1982. *The snail's spell.* New York: Penguin.

———. 1989. *Where butterflies grow.* New York: Penguin.

Sandved, Kjell B. 1999. *The butterfly alphabet.* New York: Scholastic, Inc.

Selsam, Millicent Ellis. 1981. *Where do they go? Insects in winter.* New York: Scholastic, Inc.

Selsam, Millicent Ellis, and Ronald Goor. 1991. *Backyard insects.* New York: Scholastic, Inc.

Stevens, C. 1961. *Catch a cricket.* Reading, MA: Young Scott.

Still, John. 1991. *Eyewitness Juniors: Amazing beetles.* New York: Knopf.

———. 1991. *Eyewitness Juniors: Amazing butterflies and moths.* New York: Knopf.

Watts, Barrie. 1989. *Keeping Minibeasts: Beetles.* New York: Franklin Watts.

———. 1989. *Keeping Minibeasts: Caterpillars.* New York: Franklin Watts.

———. 1991. *Keeping Minibeasts: Grasshoppers and crickets.* New York: Franklin Watts.

Wildsmith, Brian. 1993. *Look closer.* San Diego: Harcourt Brace.

Yoshi, and Ruth Wells. 1990. *The butterfly hunt.* Saxonville, MA: Picture Book Studio.

Zoehfeld, Kathleen Weidner. 1996. *Ladybug at Orchard Avenue.* Norwalk, CT: Soundprints.

## Land Use

Bang, Molly. 1997. *Common ground: The water, earth, and air we share.* New York: Scholastic, Inc.

Bartlett, Margaret. F. 1960. *The clean brook*. New York: Thomas Y. Crowell Company.

Barton, Byron. 1990. *Building a house*. New York: William Morrow.

Baylor, Byrd. 1978. *The other way to listen*. New York: Simon and Schuster.

Burton, Virginia Lee. 1978. *The little house*. Boston: Houghton Mifflin.

Carrick, Carol, and Donald Carrick. 1967. *The brook*. New York: Macmillan.

Fleming, Denise. 1996. *Where once there was a wood*. New York: Scholastic, Inc.

Gibbons, Gail. 1987. *The pottery place*. New York: Harcourt Brace.

Lyon, George Ella. 1996. *Who came down that road?* New York: Orchard Books.

McCloskey, Robert. 1976. *Blueberries for Sal*. New York: Viking Press.

Merriam, Eve. 1998. *Bam bam bam*. New York: Scholastic, Inc.

Peet, Bill. 1981. *Farewell to Shady Glade*. Boston: Houghton Mifflin.

Provensen, Alice, and Martin Provensen. 1987. *Shaker Lane*. New York: Trumpet Club

Rockwell, Anne, and Harlow Rockwell. 1992. *The first snowfall*. New York: Simon and Schuster.

Ruis, Maria, and J. M. Parramon. 1985. *The four elements: Fire*. Woodbury, NY: Barron's.

Schick, Eleanor. 1969. *City in summer*. New York: Macmillan.

Showers, Paul. 1993. *The listening walk*. New York: HarperCollins.

Tresselt, Alvin. 1990. *Wake up city!* New York: Lothrop, Lee and Shepard.

Williams, Vera B. 1984. *Three days on a river in a red canoe*. New York: William Morrow.

Yolen, Jane. 1992. *Letting Swift River go*. Boston: Little, Brown and Company.

## Mammals

Arnosky, Jim. 1986. *Deer at the brook*. New York: Lothrop, Lee and Shepard.

———. 1989. *Come out, muskrats*. New York: Lothrop, Lee and Shepard.

———. 1991. *Raccoons and ripe corn*. New York: William Morrow.

———. 1997. *Rabbits and raindrops*. New York: Scholastic, Inc.

———. 1999. *Big Jim and the white-legged moose*. New York: Lothrop, Lee and Shepard.

———. 1999. *Otters under water*. New York: Penguin Putnam.

Baker, Alan. 1990. *Two tiny mice*. New York: Scholastic, Inc.

Bancroft, Henrietta. 1997. *Animals in winter*. New York: HarperCollins.

Bland, Celia. 1996. *Bats*. New York: Scholastic, Inc.

Branley, Franklyn M. 1960. *Big tracks, little tracks*. New York: Thomas Y. Crowell.

Chase, Edith N. 1991. *The new baby calf*. New York: Scholastic, Inc.

Davies, Nicola. 2001. *Bat loves the night*. Cambridge, MA: Candlewick Press.

Dewey, Jennifer. 1989. *Can you find me? A book about animal camouflage*. New York: Scholastic, Inc.

Dorros, Arthur. 1991. *Animal tracks*. New York: Scholastic, Inc.

Ehlert, Lois. 1993. *Nuts to you!* New York: Harcourt Brace.

Fraser, Mary Ann. 1999. *Where are the night animals?* New York: HarperCollins.

Freschet, Berniece. 1978. *Possum baby*. New York: G. P. Putnam's Sons.

George, Jean Craighead. 1997. *Look to the north: A wolf pup diary*. New York: Scholastic, Inc.

Gibbons, Gail. 1987. *The milk makers*. New York: Simon and Schuster.

———. 1999. *Bats*. New York: Scholastic, Inc.

Heller, Ruth. 1999. *Animals born alive and well*. New York: Putnam Publishing.

Jarrell, Randall. 1964. *A bat is born*. Garden City, NY: Doubleday.

Jeunesse, Gallimard, and Laura Bour. 1992. *Bears: A First Discovery Book*. New York: Scholastic, Inc.

Jeunesse, Gallimard, and Sylvaine Perols. 1998. *Night creatures: A First Discovery Book*. New York: Scholastic, Inc.

Kahney, Regina. 1992. *The glow-in-the-dark book of animal skeletons*. New York: Random House.

Kohn, Bernice. 1969. *Chipmunks*. Englewood Cliffs, NJ: Prentice-Hall.

Machotka, Hana. 1991. *What neat feet!* New York: Scholastic, Inc.

MacLulich, Carolyn. 1996. *Bats*. New York: Scholastic, Inc.

Markle, Sandra. 1997. *Outside and inside bats*. New York: Simon and Schuster.

Miles, Miska. 1967. *Rabbit garden*. Boston: Little, Brown and Company.

———. 1977. *Small rabbit*. Boston: Little, Brown and Company.

Milton, Joyce. 1994. *Bats and other animals of the night*. New York: Random House.

Pallotta, Jerry. 1991. *The furry alphabet book*. Watertown, MA: Charlesbridge.

Parker, Steve. 2000. *Eyewitness Books: Skeleton*. New York: DK Publishing.

Peet, Bill. 1981. *Farewell to Shady Glade*. Boston: Houghton Mifflin Company.

Schoenherr, John. 1991. *Bear*. New York: Scholastic, Inc.

Sherro, Victoria. 1994. *Chipmunk at Hollow Tree Land*. New York: Scholastic, Inc.

Tresselt, Alvin. 1962. *The rabbit story*. New York: Lothrop, Lee and Shepard.

————. 1970. *The beaver pond*. New York: Lothrop, Lee and Shepard.

Wildsmith, Brian. 1984. *Squirrels*. New York: Oxford University Press.

Yee, Wong Herbert. 2003. *Tracks in the snow*. New York: Henry Holt.

## Plants and Flowers

Bjork, Christina, and Lena Anderson. 1988. *Linnea's windowsill garden*. New York: R and S Books.

Brown, Marc. 1981. *Your first garden book*. Boston: Little, Brown and Company.

Bunting, Eve. 2000. *Flower garden*. New York: Voyager.

Cherry, Lynne. 2003. *How groundhog's garden grew*. New York: Blue Sky Press.

Cole, Henry. 1995. *Jack's garden*. New York: Scholastic, Inc.

Ehlert, Lois. 1990. *Growing vegetable soup*. New York: Harcourt Brace.

Gibbons, Gail. 1991. *From seed to plant*. New York: Scholastic, Inc.

Hall, Zoe. 1994. *It's pumpkin time!* New York: Scholastic, Inc.

Heller, Ruth. 1999. *Plants that never bloom*. New York: Putnam Publishing.

————. 1999. *The reason for a flower*. New York: Putnam Publishing.

Hood, Susan. 1998. *National Audubon Society: First Field Guide: Wildflowers*. New York: Scholastic, Inc.

Jeunesse, Gallimard, Claude Delafosse, and René Mettler. 1991. *Flowers: A First Discovery Book*. New York: Scholastic, Inc.

Johnson, Neil. 1997. *A field of sunflowers*. New York: Scholastic, Inc.

Krauss, Ruth. 1993. *The carrot seed*. New York: HarperCollins.

————. 2000. *The growing story*. New York: HarperCollins.

McMillan, Bruce. 1995. *Counting wildflowers*. New York: William Morrow.

Steele, Mary Q. 1988. *Anna's summer songs*. New York: Greenwillow Books.

————. 1989. *Anna's garden songs*. New York: William Morrow.

Stewart, Sarah. 1997. *The gardener*. New York: Scholastic, Inc.

## Recycling and Pollution

Asch, Frank. 1994. *The earth and I are friends*. New York: Scholastic, Inc.

Berger, Melvin. 1994. *Oil spill!* New York: HarperCollins.

Cherry, Lynne. 1992. *A river ran wild: An environmental history*. New York: Harcourt Brace.

Gibbons, Gail. 1996. *Recycle: A handbook for kids*. Boston: Little, Brown and Company.

Ingpen, Robert, and Margaret Dunkle. 1994. *Conservation*. Melbourne, Australia: Hill of Content.

Peet, Bill. 1981. *Farewell to Shady Glade*. Boston: Houghton Mifflin.

————. 1981. *The Wump world*. Boston: Houghton Mifflin.

Rockwell, Harlow. 1974. *The compost heap*. Garden City, NY: Doubleday.

Showers, Paul. 1994. *Where does the garbage go?* New York: HarperCollins.

Smaridge, Norah. 1972. *Litterbugs come in every size*. Racine, WI: Golden Books.

Van Allsburg, Chris. 1990. *Just a dream*. Boston: Houghton Mifflin.

## Rocks

Barkan, Joanne. 1990. *Rocks, rocks, big and small*. Englewood Cliffs, NJ: Silver Press.

Baylor, Byrd. 1984. *If you are a hunter of fossils*. New York: Simon and Schuster.

————. 1985. *Everybody needs a rock*. New York: Simon and Schuster.

Bramwell, Martyn. 1983. *Understanding and collecting rocks and fossils*. London: Usborne.

Brandenberg, Aliki. 1990. *Fossils tell of long ago*. New York: HarperCollins.

Branley, Franklyn M. 1986. *Volcanoes*. New York: HarperCollins.

————. 1994. *Earthquakes*. New York: HarperCollins.

Challoner, Jack. 1998. *Learn about rocks and minerals*. New York: Lorenz Books.

Christian, Peggy. 2000. *If you find a rock*. New York: Harcourt, Inc.

Cole, Joanna. 1989. *The Magic School Bus: Inside the Earth*. New York: Scholastic, Inc.

Curran, Eileen. 1989. *Mountains and volcanoes*. Mahwah, NJ: Troll Associates.

Fuchshuber, Annegert. 1983. *From dinosaurs to fossils: A Start to Finish Book*. Minneapolis: Carolrhoda Books, Inc.

Gibbons, Gail. 1987. *The pottery place*. San Diego: Harcourt Brace.

Hiscock, Bruce. 1988. *The big rock*. New York: Simon and Schuster.

Hooper, Meredith. 1996. *The pebble in my pocket*. New York: Viking.

Hurst, Carol Otis. 2001. *Rocks in his head*. New York: Greenwillow Books.

Jeunesse, Gallimard, and Jean-Pierre Verdet. 1992. *The earth and sky: A First Discovery Book*. New York: Scholastic, Inc.

LaPierre, Yvette. 1994. *Native American rock art: Messages from the past*. West Palm Beach, FL: Lickle Publishing.

Lionni, Leo. 1961. *On my beach there are many pebbles*. New York: Astor-Honor Publishing.

McGovern, Ann. 1968. *Stone soup*. New York: Scholastic, Inc.

Mitgutsch, Ali. 1983. *From sand to glass: A Start to Finish Book*. Minneapolis: Carolrhoda Books.

Murphy, Stuart J. 2000. *Dave's down-to-earth rock shop*. New York: HarperCollins.

Oliver, Ray. 1993. *Hobby handbooks: Rocks and fossils*. New York: Random House.

Parker, Steve. 1997. *Eyewitness explorers: Rocks and minerals*. New York: DK Publishing.

Pellant, Chris. 2000. *The best book of fossils, rocks, and minerals*. New York: Kingfisher.

Spickert, Diane Nelson. 2000. *Earthsteps: A rock's journey through time*. Golden, CO: Fulcrum Kids.

Zoehfeld, Kathleen Weidner. 1995. *How mountains are made*. New York: HarperCollins.

## Seasons

Adolf, Arnold, and Jerry Pinkney. 1991. *In for winter, out for spring*. New York: Trumpet Club.

Ball, Jacqueline A. 1989. *What can it be? Riddles about the seasons*. Englewood Cliffs, NJ: Silver Press.

Balzola, Asun, and J. M. Parramon. 1981. *Spring*. New York: Delair Publishing.

Bjork, Christina, and Lena Anderson. 1989. *Linnea's almanac*. New York: R and S Books.

Branley, Franklyn M. 1986. *Sunshine makes the seasons*. New York: HarperCollins.

Brennan, Linda Crotta. 1997. *Flannel kisses*. New York: Houghton Mifflin.

Brown, Margaret Wise. 2002. *Give yourself to the rain: Poems for the very young*. New York: Margaret K. McElderry Books.

Bunting, Eve. 1977. *Winter's coming*. New York: Harcourt Brace.

Burningham, John. 1969. *Seasons*. New York: The Bobbs-Merrill Company.

Burton, Virginia Lee. 1974. *Katy and the big snow*. Boston: Houghton Mifflin.

———. 1978. *The little house*. Boston: Houghton Mifflin.

dePaola, Tomie. 1977. *Four stories for four seasons*. New York: Simon and Schuster.

Downing, Julie. 1991. *White snow, blue feather*. New York: Simon and Schuster.

Florian, Douglas. 1987. *A winter day*. New York: William Morrow.

Foster, D. V. 1961. *A pocketful of seasons*. New York: Lothrop, Lee and Shepard.

Franco, Betsy. 1994. *Fresh fall leaves*. New York: Scholastic, Inc.

Gomi, Taro. 1989. *Spring is here*. San Francisco: Chronicle Books.

Hall, Zoe. 2000. *Fall leaves*. New York: Scholastic, Inc.

Hirschi, Ron. 1990. *Spring*. New York: Scholastic, Inc.

Keats, Ezra Jack. 1981. *The snowy day*. New York: Viking Press.

Lenski, Lois. 1953. *On a summer day*. New York: Henry Z. Walck.

———. 1973. *Spring is here*. New York: Henry Z. Walck.

———. 2000. *I like winter*. New York: Random House.

———. 2000. *Now it's fall*. New York: Random House.

Locker, Thomas. 1995. *Seeing science through art: Sky tree*. New York: HarperCollins.

Maass, Robert. 1993. *When winter comes*. New York: Scholastic, Inc.

———. 1994. *When spring comes*. New York: Scholastic, Inc.

Maestro, Betsy. 1989. *Snow day*. New York: Scholastic, Inc.

Parramon, J. M., and Ulises Wensel. 1981. *Autumn*. New York: Delair Publishing.

Provensen, Alice. 1978. *The year at Maple Hill Farm*. New York: Atheneum.

Rockwell, Anne. 1996. *My spring robin*. New York: Simon and Schuster.

Rogasky, Barbara, ed. 1994. *Winter poems*. New York: Scholastic, Inc.

———. 2001. *Leaf by leaf: Autumn poems*. New York: Scholastic, Inc.

Rylant, Cynthia. 2000. *In November*. San Diego: Harcourt.

Sams, Carl R., II, and Jean Stoick. 2000. *Stranger in the woods: A photographic fantasy*. Milford, MI: Carl R. Sams II Photography.

Schick, Eleanor. 1970. *City in the winter*. New York: Macmillan.

Senisi, Ellen B. 2001. *Fall changes.* New York: Scholastic, Inc.

Tresselt, Alvin. 1950. *Hi, mister robin!* New York: Lothrop, Lee and Shepard.

———. 1989. *White snow, bright snow.* New York: William Morrow.

Tudor, Tasha. 1957. *Around the year.* New York: Henry Z. Walck.

Updike, John. 1999. *A child's calendar.* New York: Scholastic, Inc.

Wildsmith, Brian. 1991. *Animal season.* New York: Oxford University Press.

## Seeds

Anthony, Joseph, and Cris Arbo. 1997. *The dandelion seed.* Nevada City, CA: Dawn Publications.

Berger, Melvin. 1992. *All about seeds.* New York: Scholastic, Inc.

Bjork, Christina, and Lena Anderson. 1988. *Linnea's windowsill garden.* New York: R and S Books.

Brandenberg, Aliki. 1976. *Corn is maize: The gift of the Indians.* New York: HarperCollins.

Carle, Eric. 1998. *The tiny seed.* New York: Simon and Schuster.

Ehlert, Lois. 1990. *Growing vegetable soup.* New York: Harcourt Brace.

———. 1992. *Planting a rainbow.* New York: Harcourt Brace.

———. 1993. *Eating the alphabet.* New York: Harcourt Brace.

Gibbons, Gail. 1991. *From seed to plant.* New York: Scholastic, Inc.

———. 2000. *Apples.* New York: Scholastic, Inc.

Givens, Janet. 1982. *Something wonderful happened.* New York: Atheneum.

Hall, Zoe. 1994. *It's pumpkin time!* New York: Scholastic, Inc.

Heller, Ruth. 1984. *Plants that never bloom.* New York: Grosset and Dunlap.

———. 1999. *The reason for a flower.* New York: Putnam Publishing.

Jeunesse, Gallimard, and Pascale de Bourgoing. 1991. *Fruit: A First Discovery Book.* New York: Scholastic, Inc.

Jordan, Helene J. 1962. *Seeds by wind and water.* New York: Thomas Y. Crowell.

———. 1992. *How a seed grows.* New York: HarperCollins.

Merrill, Claire. 1973. *A seed is a promise.* New York: Scholastic, Inc.

Petie, Haris. 1976. *The seed the squirrel dropped.* Englewood Cliffs, NJ: Prentice-Hall.

Selsam, Millicent Ellis. 1959. *Seeds and more seeds.* New York: HarperCollins.

Titherington, Jeanne. 1990. *Pumpkin pumpkin.* New York: William Morrow.

## Shadows

Bulla, Clyde Robert. 1994. *What makes a shadow?* New York: HarperCollins.

DeRegniers, Beatrice Schenk. 1960. *The shadow book.* New York: Harcourt Brace.

Dodd, Anne Wescott. 1992. *Footprints and shadows.* New York: Simon and Schuster.

Dorros, Arthur. 1990. *Me and my shadow.* New York: Scholastic, Inc.

Gore, Shelia. 1989. *My shadow.* New York: Doubleday.

Otto, Carolyn G. 2001. *Shadows.* New York: Scholastic, Inc.

Paul, Ann Whitford. 1992. *Shadows are about.* New York: Scholastic, Inc.

Stevenson, Robert Louis. 1999. *My shadow.* New York: Candlewick Press.

Swinburne, Stephen R. 1999. *Guess whose shadow?* Honesdale, Pa.: Boyds Mill Press.

## Soil

Dunrea, Oliver. 1989. *Deep down underground.* New York: Macmillan.

Gibbons, Gail. 1987. *The pottery place.* New York: Harcourt Brace.

Glaser, Linda. 1992. *Wonderful worms.* Brookfield, CT: Millbrook Press.

Hawkinson, John. 1965. *The old stump.* Chicago: Albert Whitman and Company.

Knutson, Kimberly. 1992. *Muddigush.* New York: Macmillan.

Pfeffer, Wendy. 2004. *Wiggling worms at work.* New York: HarperCollins.

Rockwell, Harlow. 1974. *The compost heap.* Garden City, NY: Doubleday.

Romanova, Natalia. 1992. *Once there was a tree.* New York: Penguin USA.

Ryder, Joanna. 1990. *Under your feet.* New York: Four Winds Press.

Tresselt, Alvin. 1972. *The dead tree.* New York: *Parents'* Magazine Press.

Wong, Herbert H. 1977. *Our earthworms.* Reading, PA: Addison-Wesley.

## Spiders

Back, Christine. 1984. *Spider's web.* Morristown, NJ: Silver Burdett.

Canizares, Susan. 1998. *Spider names.* New York: Scholastic, Inc.

Chinery, Michael. 1991. *Spider*. Mahwah, NJ: Troll Associates.

Freschet, Berniece. 1972. *The web in the grass*. New York: Charles Scribner's Sons.

Gibbons, Gail. 1994. *Spiders*. New York: Holiday House.

Graham, Margaret Bloom. 1967. *Be nice to spiders*. New York: HarperCollins.

Markel, Sandra. 1994. *Outside and inside spiders*. New York: Simon and Schuster.

Parker, John. 1988. *I love spiders*. New York: Scholastic, Inc.

Parsons, Alexandra. 1990. *Eyewitness Juniors: Amazing spiders*. New York: Knopf.

Petty, Kate. 1985. *Spiders*. New York: Aladdin Books.

Winner, Yvonne. 1996. *Spiders spin webs*. Cambridge, MA: Charlesbridge.

## Trees and Forests

Behn, Harry. 1992. *Trees*. New York: Henry Holt.

Canizares, Susan. 1998. *Evergreens are green*. New York: Scholastic, Inc.

Canizares, Susan, and Pamela Chanko. 1998. *Look at this tree*. New York: Scholastic, Inc.

Canizares, Susan, and Daniel Moreton. 1998. *Who lives in a tree?* New York: Scholastic, Inc.

Canizares, Susan, and Mary Reid. 1998. *Treats from a tree*. New York: Scholastic, Inc.

Cassie, Brian, and Marjorie Burns. 1999. *National Audubon Society: First Field Guide: Trees*. New York: Scholastic, Inc.

Cristini, Ermanno, and Luigi Puricelli. 1983. *In the woods*. New York: Scholastic, Inc.

Curran, Eileen. 1989. *Look at a tree*. Mahwah, NJ: Troll Associates.

Davis, Wendy. 1995. *From tree to paper*. New York: Scholastic, Inc.

Davol, Marguerite W. 1992. *The heart of the wood*. New York: Simon and Schuster.

de Bourgoing, Pascale. 1992. *The tree: A First Discovery Book*. New York: Scholastic, Inc.

Ehlert, Lois. 1991. *Red leaf, yellow leaf*. New York: Harcourt Brace.

Ernst, Kathryn. 1976. *Mr. Tamarin's trees*. New York: Crown Publishers.

Fleming, Denise. 2000. *Where once there was a wood*. New York: Henry Holt.

Florian, Douglas. 1991. *A carpenter*. New York: Greenwillow Books.

Fowler, Allan. 1991. *It could still be a tree*. Chicago: Children's Press.

Gackenbach, Dick. 1992. *Mighty tree*. San Diego: Harcourt Brace.

Gibbons, Gail. 2002. *Tell me, tree: All about trees for kids*. New York: Little, Brown and Company.

Hawkinson, John. 1965. *The old stump*. Chicago: Albert Whitman and Company.

Hickman, Pamela. 1999. *Starting with nature: Tree book*. New York: Kids Can Press.

Knutson, Kimberly. 1993. *Ska-tat!* New York: Macmillan.

Lauber, Patricia. 1999. *Be friends to trees*. New York: HarperCollins.

Maestro, Betsy. 1994. *Why do leaves change colors?* New York: HarperCollins.

Miklowitz, Gloria D. 1998. *Save that raccoon!* New York: Harcourt Brace.

Miles, Miska. 1966. *Fox and fire*. Boston: Little, Brown and Company.

Oppenheim, Joanne. 1967. *Have you seen trees?* Chicago: Children's Press.

Pondendorf, Illa. 1954. *The true book of trees*. Chicago: Children's Press.

Quinn, Grey Henry. 1994. *A gift of a tree*. New York: Scholastic, Inc.

Reed-Jones, Carol. 1995. *The tree in the ancient forest*. Nevada City, CA: Dawn Publications.

Robbins, Ken. 1998. *Autumn leaves*. New York: Scholastic, Inc.

Rogasky, Barbara. 2001. *Leaf by leaf: Autumn poems*. New York: Scholastic, Inc.

Romanova, Natalia. 1992. *Once there was a tree*. New York: Penguin USA.

Schreiber, Anne. 1994. *Log hotel*. New York: Scholastic, Inc.

Selsam, Millicent Ellis. 1968. *Maple tree*. New York: William Morrow.

Tresselt, Alvin. 1948. *Johnny maple-leaf*. New York: Lothrop, Lee and Shepard.

———. 1972. *The dead tree*. New York: *Parents'* Magazine Press.

Udry, Janice May. 1987. *A tree is nice*. New York: HarperCollins.

## Weather

Ariane. 1996. *Small cloud*. New York: Walker and Company.

Bacon, Ron. 1993. *Wind*. New York: Scholastic, Inc.

Barrett, Judi. 1977. *The wind thief*. New York: Atheneum.

Branley, Franklyn M. 1988. *Tornado alert*. New York: HarperCollins.

———. 1997. *Down comes the rain*. New York: HarperCollins.

———. 1999. *Flash, crash, rumble, and roll*. New York: HarperCollins.

———. 2000. *Snow is falling*. New York: HarperCollins.

Canizares, Susan, and Betsey Chessen. 1998. *Storms*. New York: Scholastic, Inc.

———. 1998. *Wind*. New York: Scholastic, Inc.

Canizares, Susan, and Daniel Moreton. 1998. *Sun*. New York: Scholastic, Inc.

Chaffin, Lillie D. 1963. *Bear weather*. New York: Macmillan.

Chanko, Pamela, and Daniel Moreton. 1998. *Weather*. New York: Scholastic, Inc.

de Bourgoing, Pascale. 1991. *Weather: A First Discovery Book*. New York: Scholastic, Inc.

dePaola, Tomie. 1985. *The cloud book*. New York: Holiday House.

DeWitt, Lynda. 1993. *What will the weather be?* New York: HarperCollins.

Ehlert, Lois. 1995. *Snowballs*. San Diego: Harcourt Brace.

Freeman, Don. 1978. *A rainbow of my own*. New York: Viking Press.

Garelick, May. 1997. *Where does the butterfly go when it rains?* Greenvale, NY: Mondo Publishing.

Gibbons, Gail. 1990. *Weather words and what they mean*. New York: Holiday House.

Hutchins, Pat. 1993. *The wind blew*. New York: Simon and Schuster.

Kahl, Jonathan. 1998. *National Audubon Society: First Field Guide: Weather*. New York: Scholastic, Inc.

Kalan, Robert. 1991. *Rain*. New York: William Morrow.

Lindberg, Reeve. 1996. *What is the sun?* Cambridge, MA: Candlewick Press.

Lyon, George Ella. 1990. *Come a tide*. New York: Orchard Books.

McCloskey, Robert. 1989. *Time of wonder*. New York: Viking Press.

Polacco, Patricia. 1990. *Thundercake*. New York: Philomel Books.

Rockwell, Anne, and Harlow Rockwell. 1992. *The first snowfall*. New York: Simon and Schuster.

Rogers, Paul. 1989. *What will the weather be like today?* New York: Greenwillow Books.

Schaefer, Lola M. 2001. *This is rain*. New York: HarperCollins.

Serfozo, Mary. 1993. *Rain talk*. New York: Simon and Schuster.

Shaw, Charles G. 1993. *It looked like spilt milk*. New York: HarperCollins.

Spier, Peter. 1997. *Peter Spier's rain*. New York: Doubleday.

Shulevitz, Uri. 1998. *Snow*. New York: Scholastic, Inc.

Tresselt, Alvin. 1946. *Rain drop splash*. New York: Lothrop, Lee and Shepard.

———. 1959. *Follow the wind*. New York: Lothrop, Lee and Shepard.

———. 1988. *Hide and seek fog*. New York: William Morrow.

Yashima, Taro. 1985. *Umbrella*. New York: Penguin Putnam.

Zolotow, Charlotte. 1989. *The storm book*. New York: HarperCollins.

# Appendix B

# Teacher Resources

## Books and Guides

Anderson, Alan, Gwen Diehn, and Terry Krautwurst. 1998. *Geology crafts for kids: Fifty nifty projects to explore the marvels of planet Earth.* New York: Sterling Publishing.

Blobaum, Cindy. 1999. *Geology rocks! Fifty hands-on activites to explore the earth.* Charlotte, VT: Williamson Publishing Co.

Carlson, Laurie, and Judith Dammel. 1995. *Kids camp! Activities for the backyard or wilderness.* Chicago: Chicago Review Press.

Carmichael, Viola S. 1969. *Science experiences for young children.* Sierra Madre, CA: SCAEYC.

Cohen, Richard, and Betty Phillips Tunick. 1997. *Snail trails and tadpole tails: Nature education guide for young children.* St. Paul: Redleaf Press.

Cornell, Joseph. 1998. *Sharing nature with children.* Nevada City, CA: Dawn Publications.

Diehn, Gwen, and Terry Krautwurst. 1997. *Kid style: Nature crafts: 50 terrific things to make with nature's materials.* New York: Sterling Publishing.

———. 1997. *Science crafts: Fifty fantastic things to invent and create.* New York: Sterling Publishing.

Diener, Carolyn S., C. R. Jettinghoff, E. B. Robertson, and M. P. Stickland. 1982. *Energy: A curriculum unit for three-, four-, and five-year-olds.* Atlanta: Humanics Limited.

Drake, Jane, and Ann Love. 1994. *The kids' summer handbook.* Tonawanda, NY: Kids Can Press.

———. 1998. *The kids' campfire book.* Tonawanda, NY: Kids Can Press.

Drinkard, Lawson. 1999. *Fishing in a brook: Angling activities for kids.* Layton, UT: Gibbs Smith.

Fiarotta, Phyllis. 1975. *Snips and snails and walnut shells: Nature crafts for children.* New York: Workman Publishing Company.

Gertz, Susan E., Dwight J. Portman, and Mickey Sarquis. 1996. *Teaching physical science through children's literature.* Middletown, OH: Terrific Science Press.

Hickman, Pamela M. 1988. *Birdwise: Forty fun feats for finding out about our feathered friends.* Reading, MA: Addison-Wesley Publishing.

———. 1990. *Bugwise: Thirty incredible insect investigations and arachnid activities.* Reading, MA: Addison-Wesley Publishing.

Horsfall, Jacqueline. 1997. *Play lightly on the earth: Nature activities for children three to nine years old.* Nevada City, CA: Dawn Publications.

Katz, Adrienne. 1986. *Naturewatch.* Reading, MA: Addison-Wesley.

Kohl, MaryAnn F. 1992. *Mudworks: Creative clay, dough, and modeling experiences.* Bellingham, WA: Bright Ring Publishing.

Kohl, MaryAnn F., and Cindy Gainer. 1991. *Good earth art: Environmental art for kids.* Bellingham, WA: Bright Ring Publishing.

Kohl, MaryAnn F., and Jean Potter. 1993. *Science arts: Discovering science through art experiences.* Bellingham, WA: Bright Ring Publishing.

Kraul, Walter. 1995. *Earth, water, fire, and air: Playful explorations in the four elements.* Mt. Rainier, MD: Gryphon House.

Lawton, Rebecca, Diane Lawton, and Susan Panttaja. 1997. *Discover nature in the rocks: Things to know and things to do.* Mechanicsburg, PA: Stackpole Books.

Linglebach, Jenepher, ed. 1989. *Hands-on nature: Information and activities for exploring the environment with children.* Woodstock, VT: Vermont Institute of Nature Science.

Milord, Susan. 1996. *The kids' nature book: 365 indoor/outdoor activities and experiences.* Charlotte, VT: Williamson Publishing.

Mitchell, John. 1996. *The curious naturalist.* Lincoln, MA: Massachusetts Audubon Society.

National Wildlife Federation. 1997. *Ranger Rick's Naturescope: Astronomy adventures.* New York: McGraw-Hill.

———. 1997. *Ranger Rick's Naturescope: Birds, birds, birds!* New York: McGraw-Hill.

———. 1997. *Ranger Rick's Naturescope: Endangered species: Wild and rare.* New York: McGraw-Hill.

———. 1997. *Ranger Rick's Naturescope: Geology: The active earth.* New York: McGraw-Hill.

———. 1997. *Ranger Rick's Naturescope: Let's hear it for herps.* New York: McGraw-Hill.

———. 1997. *Ranger Rick's Naturescope: Wading into wetlands.* New York: McGraw-Hill.

———. 1997. *Ranger Rick's Naturescope: Wild about weather.* New York: McGraw-Hill.

———. 1998. *Ranger Rick's Naturescope: Amazing mammals: Part 1.* New York: McGraw-Hill.

———. 1998. *Ranger Rick's Naturescope: Amazing mammals: Part 2.* New York: McGraw-Hill.

———. 1998. *Ranger Rick's Naturescope: Incredible insects.* New York: McGraw-Hill.

———. 1998. *Ranger Rick's Naturescope: Pollution: Problems and solutions.* New York: McGraw-Hill.

———. 1988. *Ranger Rick's Naturescope: Trees are terrific!* New York: McGraw-Hill.

Needham, Bobbe. 1998. *Ecology crafts for kids: Fifty great ways to make friends with planet Earth.* New York: Sterling Publishing Company.

Nickelsburg, Janet. 1976. *Nature activities for early childhood.* Reading, MA: Addison-Wesley.

Rhoades, Diane. 1998. *Garden crafts for kids: Fifty great reasons to get your hands dirty.* New York: Sterling Publishing Company.

Sisson, Edith A. 1982. *Nature with children of all ages: Activities and adventures for exploring, learning, and enjoying the world around us.* Englewood Cliffs, NJ: Prentice-Hall, Inc.

Suzuki, David. 1992. *Looking at insects.* New York: John Wiley and Sons.

VanCleave, Janice. 1996. *Rocks and minerals: Mind-boggling experiments you can turn into science fair projects.* New York: John Wiley and Sons.

———. 1998. *Insects and spiders: Mind-boggling experiments you can turn into science fair projects.* New York: John Wiley and Sons.

———. 2000. *Science around the year.* New York: John Wiley and Sons.

White, Linda. 1996. *Cooking on a stick: Campfire recipes for kids.* Layton, UT: Gibbs Smith.

———. 1998. *Sleeping in a sack: Camping activities for kids.* Layton, UT: Gibbs Smith.

———. 2000. *Trekking on a trail: Hiking adventures for kids.* Layton, UT: Gibbs Smith.

## Organizations and Resources

American Sportfishing Association
1033 North Fairfax Street #200
Alexandria, VA 22314
703-519-9691
www.asafishing.org

Alliance to Save Energy
1200-18th Street NW, Suite 900
Washington, DC 20036
202-857-0666
www.ase.org

American Fisheries Society
5410 Grosvenor Lane, Suite 110
Bethesda, MD 20814
301-897-8616
www.fisheries.org

Co-op America
1612 K Street NW
Washington, DC 20006
800-58-GREEN
www.coopamerica.org

Forestkeepers
4207 Lindell Boulevard, Suite 120
St. Louis, MO 63108
800-9-FOREST
www.forestkeepers.org

Groundwater Foundation
P. O. Box 22558
Lincoln, NE 68542-2558
800-858-4844
www.groundwater.org

Missouri Department of Conservation
P. O. Box 180
Jefferson City, MO 65201-0180
573-751-4115
www.conservation.state.mo.us

National Audubon Society
700 Broadway
New York, NY 10003
212-979-3000
www.audubon.org

National Trappers Association, Inc.
4111 East Starr Avenue
Nacogdoches, TX 75961
(no phone known)
www.nationaltrappers.com

National Wildlife Federation
8925 Leesburg Pike
Vienna, VA 22180
703-790-4000
www.nwf.org

North American Association for Environmental Education
410 Tarvin Road
Rock Spring, GA 30739
706-764-2708
www.naaee.org

Project Learning Tree
www.plt.org
An environmental education program available through the
American Forest Foundation
1111-19th Street NW, Suite 780
Washington, DC 20036
202-463-2462

Stream Team
Missouri Department of Conservation
P.O. Box 180
Jefferson City, MO 65102
573-751-4115
www.mostreamteam.org

Trees for Life
3006 West St. Louis
Wichita, KS 67203
316-945-6929
www.treesforlife.org

USDA Forest Service
1400 Independence Avenue SW
Washington, DC 20250-0003
202-205-8333
www.fs.fed.us

U.S. Department of Energy
1000 Independence Avenue SW
Washington, DC 20585
202-586-5575
www.energy.gov

U.S. Environmental Protection Agency
1200 Pennsylvania Avenue NW
Washington, DC 20460
202-272-0167
www.epa.gov

U.S. Fish and Wildlife Service
Department of the Interior
1849 C Street NW
Washington, DC 20240
800-344-WILD
www.fws.gov

World Wildlife Fund
1250-24th Street NW
P. O. Box 97180
Washington, DC 20037
800-CALL-WWF
www.worldwildlife.org

# APPENDIX C

## Lesson Plans

## Air and Water

Concepts children might construct:
>Air and water are important to all living things.
>
>Water cycles over and over again.
>
>Water cycles to earth in several different states (solid: snow, liquid: rain, and gas: fog).
>
>Air is difficult to see but can be felt and heard.
>
>People can affect both water and air quality.

### Art

Ice sculptures W-10 ✻
What Goes in the Wind? SP-4
Wind chimes SP-4 ✻
Blow painting SP-4 ✻
Aquatic life pictures SU-9 ✻

### Blocks/Pretend Play

Ice Fishing W-10 ✻
Rain Gear SP-1 ✻
Hanging Out SP-2
Build shelters for animals to go in rain SP-3 ✻
Oil and water clean-up SU-11 ✻

### Bulletin Board

Display pictures of how wind helps and harms people SP-4 ✻

### Field Trips

Walk in rain SP-1 ✻
Look for signs of erosion SP-9 ✻
Aquatic Life SU-9

### Group

What Is Erosion? SP-9

### Large Motor/Outside

Experiment with wind traveling seeds F-11 ✻
Look for seeds that travel with the wind F-11 ✻
Skaters Away W-10

Snow sculptures W-10 ✻
Let's measure the rain SP-1 ✻
Pretend to be raindrops SP-1 ✻
Water painting SP-2 ✻
Crepe paper streamers SP-4 ✻
Kites SP-4 ✻
Blow bubbles SP-4 ✻
Feel air SU-7 ✻
Move like animals in water SU-9 ✻

### Manipulative

Where Do Animals Go When It Rains? SP-3

### Music

I'm a Little Milkweed Cradle F-11
Three rain drops SP-1 ✻
Scarf dance in wind SP-4 ✻

### Science

Do trees get drinks? F-5
Observe a puddle SP-2 ✻
Chart wind creations SP-4 ✻
What Is Air? SU-7
Bubble prints SU-7 ✻
What's in the Air? SU-8
Water samples SU-9 ✻
Underwater feely box SU-9 ✻

### Stories/Writing

B, T, and U W-13
Drip and Drop SP-1
Joe's Choice SU-10

## Amphibians and Reptiles

Concepts children might construct:
>Amphibians go through metamorphosis, reptiles do not.
>
>Amphibians and reptiles are cold blooded.

✻ indicates supplementary activity

Amphibians lay eggs, usually in water.
Reptiles hatch from eggs, usually buried on land.
Frogs and toads are different.
Amphibians and reptiles are both predator and prey.
Amphibians and reptiles live on land and in water.
Reptiles have scales, amphibians do not.
Reptiles are not slimy or scary.
Reptiles and amphibians can be identified by their
  eggs and color pattern.

## Art

Cave Life F-16
What Hatches from an Egg? SP-19
Aquatic life pictures SU-9 🎄
Fishing Permits SU-11
Frog and toad finger paint SU-13 🎄
Frog and toad habitat SU-13 🎄
Camouflage painting SU-15 🎄

## Blocks/Pretend Play

Conservation agent dress-up clothes SU-11 🎄
Frogging SU-12 🎄

## Field Trips

Aquatic Life SU-9

## Large Motor/Outside

Act out metamorphosis of frog SU-13 🎄
You Can't Find Me! SU-15

## Manipulative

Where Do Animals Go When It Rains? SP-3
Egg memory SP-19 🎄
Frog and toad sort SU-13 🎄
Sort animal pictures by skin covering SU-15 🎄

## Music

Some have long legs SU-13 🎄
"Ribbit Ribbit" SU-13 🎄

## Nutrition

Edible frogs and toads SU-14 🎄
Peanut Butter Playdough Insects SU-2 🎄

## Science

Underwater "feely" box SU-9 🎄
Observe frog and toad SU-13 🎄

## Stories/Writing

Frogs and Toads SU-13

# Aquatic life

Concepts children might construct:
  Water in streams, rivers, creeks, and lakes supports a
    variety of plant and animal life.
  Water is a type of habitat.
  Some animals can live only in water while others can
    adapt to both land and water.

## Art

Cave Life F-16
What Hatches from an Egg? SP-19
Aquatic life watercolor wash SU-9 🎄
Fishing Permits SU-11
Frog habitats SU-13 🎄
Frog and toad finger paint SU-13 🎄
Camouflage painting SU-15 🎄

## Blocks/Pretend Play

Ice fishing W-10 🎄
Oil and water clean-up SU-8 🎄
Beaver dams SU-9 🎄
Conservation agent dress-up clothes SU-10 🎄
Let's Go Fishing SU-12
Frogging SU-12 🎄

## Field Trips

Aquatic Life SU-9

## Large Motor/Outside

Move like aquatic life SU-9 🎄
Play "fish and worms" SU-12 🎄
You Can't Find Me! SU-15

## Manipulative

Where Do Animals Go When It Rains? SP-3
Tackle box sort SU-12 🎄
Frog and toad sort SU-13 🎄
Sort animals pictures by skin covering SU-15 🎄

## Music

Some have long legs SU-13 🎄
"Ribbit Ribbit" SU-13 🎄

🎄 indicates supplementary activity

## Nutrition

Peanut Butter Playdough Insects SU-2 ✀
Sardine tasting SU-12 ✀
Edible frogs and toads SU-13 ✀

## Science

Water samples SU-9 ✀
Underwater "feely" box SU-9 ✀

## Stories/Writing

Joe's choice SU-10
Frogs and Toads SU-13

# Birds

Concepts children might construct:
  Birds have feathers.
  Birds usually fly.
  Birds have wings.
  Birds hatch from eggs.
  Birds can be identified by their call, nest, eggs, and
    appearance.
  Birds are warm-blooded.
  Birds usually care for their young.

## Art

Stuffed Birds W-4
Sculpt birds W-4 ✀
Record bird song through art W-6 ✀
What Hatches from an Egg? SP-19
Makes nests SP-19 ✀
Hunting permits SU-10 ✀

## Blocks/Pretend Play

Pretend to be birds W-3 ✀
Bird wings W-4 ✀
Binoculars W-6 ✀
Conservation agent dress-up clothes SU-10 ✀

## Bulletin Board

Bird sightings W-1 ✀
Beak, feet, foods, and habitat display W-3 ✀
Hang stuffed birds W-4 ✀
Documentation of feather exploration W-5 ✀

## Field Trips

Turkey check station F-19 ✀
Turkey farm F-20 ✀

Bird hatchery SP-19 ✀
Aquatic Life SU-9

## Group

Invite a turkey hunter F-20 ✀
Chart bird feeder observations W-1 ✀
Mother bird SP-19 ✀

## Large Motor/Outside

Collect seeds and berries for food F-10 ✀
Play "catch a turkey" F-20 ✀
Put out several different types of feeders W-1 ✀
Migration Obstacles W-7
You Can't Find Me! SU-15

## Manipulative

Tree Puzzle F-6
Measure, weigh and pour birdseed W-1 ✀
Pick a Beak W-2
Sort birds by food W-2 ✀
Bird Puzzles W-3
Track Puzzlers W-8
Where Do Animals Go When It Rains? SP-3
Sort animal pictures by those who hatch from eggs
  and those who don't SP-20 ✀
Egg memory SP-19 ✀
Bird nest match SP-19 ✀
Sort animal pictures by skin coverings SU-15 ✀

## Music

I'm a Very Fine Turkey F-20
I'm a Cardinal W-6
Bird calls audiotape or CD W-6 ✀

## Nutrition

Rice cereal and eat like birds W-2 ✀
Bird nests SP-19 ✀
Gummy worms SP-19 ✀
Peanut Butter Playdough Insects SU-2 ✀

## Science

Explore wild and domestic turkey feathers F-20 ✀
Birdfeeders W-1
Take home bird feeders and journals W-1 ✀
Plant birdseed W-1 ✀
Explore a Feather W-5
Display eggshells SP-19 ✀

### Stories/Writing

Molly and the Forest Fire F-3
See How the Turkey Grows F-19
Blank books for bird stories W-4 ✀
Seeds, Roots, and Plants SP-14
Frogs and Toads SU-13

### Woodworking

Build birdhouses or birdfeeders W-1 ✀

# Camping

Concepts children might construct:

Camping is a fun way to enjoy our natural resources.
There are rules that people need to follow to be safe and to respect other campers and the wildlife.
Camping is a way to discover and explore the natural world.
Camping involves gathering specific equipment.

### Art

Forest fire pictures F-3 ✀
Sponge paint bats F-17 ✀
Flower rubbings SP-13 ✀
Noise makers SP-14 ✀
Litter bags SU-6 ✀
Fishing permits SU-11 ✀

### Blocks/Pretend Play

Binoculars W-6 ✀
Hanging Out SP-2
Let's Go Camping! SP-15
Conservation agent dress-ups SU-11 ✀
Let's Go Fishing SU-12

### Field Trips

Campfire in natural area F-3 ✀
Bird call listening W-6 ✀
Natural camping area and commercial camping area SP-14 ✀
Spider sniffing SU-1 ✀
Trash Pickup SU-6
Aquatic Life SU-9
Rock Collection SU-17

### Group

Invite a forester or Smokey the Bear to visit F-3 ✀
Invite outdoor experts to light campfire with flint SP-14 ✀
Invite an RV owner to visit SP-14 ✀
Create a first-aid kit SP-14 ✀

### Large Motor/Outside

Tree Skin F-2
Leaf rubbing scavenger hunt F-2 ✀
Pretend campfire with wood and buckets F-3 ✀
Blindfolded obstacle course F-17 ✀
Listen for bird calls W-6 ✀
Campfire safety SP-14 ✀
Compasses SP-14 ✀
Ant Café SU-4
Boat on play yard SU-12 ✀
Practice casting SU-12 ✀
Foxy Predators SU-14

### Manipulative

Track Puzzlers W-8
Where Do Animals Go When It Rains? SP-3
Flower match SP-12 ✀
Shoelaces to practice knots SP-14 ✀

### Music

"Hairy not scary" F-17 ✀
I'm a Little Fox Squirrel F-18
Bird calls audiotape or CD W-6 ✀
The lights go on SP-3 ✀
Who Am I? SP-18

### Nutrition

Trail mix SP-14 ✀

### Science

Display charred wood F-3 ✀
Worth Their Weight F-17

### Stories/Writing

Molly and the Forest Fire F-3
Map making SP-15 ✀
Joe's Choice SU-10

### Woodworking

Flower press SP-13 ✀

✀ indicates supplementary activity

# Caves

Concepts children might construct:

>Caves support a variety of wildlife.
>
>People and most mammals live in the twilight zone of caves.
>
>Different areas of the cave provide different types of habitat.
>
>Caves are made of rock.

## Art

Cave Life F-16
Cave models F-16 ✄
Sponge paint bats F-17 ✄

## Blocks/Pretend Play

Bicycle helmet "caving hat" with flashlights taped on top F-16
Hibernation Cave W-9 ✄

## Field Trips

Cave F-16 ✄
Aquatic Life SU-9
Rock Collection SU-17

## Group

Invite a caver F-16 ✄
Echo experiments F-17 ✄
Crystal garden SU-18 ✄

## Large Motor/Outside

Blindfolded obstacle course F-17 ✄

## Manipulative

Sort animal pictures by habitats F-16 ✄
Where Do Animals Go When It Rains? SP-3

## Music

"Hairy not scary" F-17 ✄

## Nutrition

Bat shaped fruit leather F-17 ✄

## Science

Worth Their Weight F-17

# Energy

Concepts children might construct:

>Energy can be created in a variety of ways.
>
>Energy is important for survival.
>
>Energy should be used wisely.

## Art

Energy collages W-13 ✄
Sun pictures W-14 ✄
Warm and cool color collages W-14 ✄
Discuss recycled products used in art W-17 ✄
Fabric collages W-20 ✄
What Goes in the Wind? SP-4
Wind chimes SP-4 ✄
Blow painting SP-4 ✄

## Blocks/Pretend Play

Refueling station for cars and trucks W-13 ✄
Grocery Shopping W-17
Cotton for insulating W-19 ✄
Hanging Out SP-2

## Bulletin Board

Pictures of people using energy W-14 ✄
Display pictures of how the wind helps people or how it can do harm SP-4 ✄

## Field Trips

Power plant W-12 ✄
Service station W-12 ✄
Wood stove store W-15 ✄
Grocery store W-15 ✄
Building site being insulated W-19 ✄

## Group

What Is Energy? W-12
Discuss colors of animal coats W-14 ✄
No Electricity W-15
Saving Energy W-16

## Large Motor/Outside

Refueling station for tricycles W-13 ✄
Snow painting W-14 ✄
Touch light and dark cars on sunny day W-14 ✄
Water paint SP-2 ✄
Crepe paper streamers SP-4 ✄
Fly kite SP-4 ✄
Shadow hunt SP-5

## Music

"Row, row, row your boat" W-13 ✻
Scarf dance in the wind SP-4 ✻

## Nutrition

Discuss refueling energy as children eat snack or lunch
   W-12 ✻
Sun tea W-13 ✻
Sun dried fruit W-14 ✻

## Science

Static electricity experiments W-12 ✻
Place plants in sunny and dark rooms W-13 ✻
Wind up toys W-13 ✻
Does the Sun Give Us Energy? W-14
Insulation W-19
Display insulation and calking W-19 ✻
Winter Fabrics W-20
Chart wind creations SP-4 ✻

## Stories/Writing

B, T, and U W-13
Fred's Forest W-18

## Woodworking

Dismantle appliances W-12 ✻

# Fish

Concepts children might construct:
   Fish need a water habitat.
   Scales cover fish bodies.
   Fish breathe with gills.
   Fish eyes are usually on the sides of the head.
   Fish have fins.
   Fish lay eggs in water.
   Fish are both predator and prey.

## Art

Cave Life F-16
What Hatches from an Egg? SP-19
Fish watercolor wash SU-9 ✻
Fishing Permits SU-11
Camouflage paint SU-15 ✻

## Blocks/Pretend Play

Ice fishing W-10 ✻
Oil and water clean-up SU-8 ✻

Conservation agent dress-up clothes SU-10 ✻
Let's Go Fishing SU-12
Materials to cook fish SU-12 ✻

## Field Trips

Aquatic Life SU-9
Fishing SU-10 ✻

## Large Motor/Outside

Move like fish SU-9 ✻
Practice casting SU-12 ✻
Play "fish and worms" SU-12 ✻
Put a boat in the play yard SU-12 ✻
You Can't Find Me! SU-15

## Manipulative

Sort animal pictures by those who hatch from eggs
   and those who don't SP-19 ✻
Egg memory SP-19 ✻
Tackle box sort SU-12 ✻
Sort animal pictures by skin covering SU-15 ✻

## Nutrition

Peanut Butter Playdough Insects SU-2 ✻
Sardine tasting SU-12 ✻

## Science

Underwater "feely" box SU-9 ✻
Water samples SU-9 ✻
Observe fish SU-12 ✻

## Stories/Writing

Joe's Choice SU-10
Fishing stories SU-10 ✻
Frogs and Toads SU-13

# Food chains

Concepts children might construct:
   People and animals get their food from plants and
      other animals.
   Predators look for prey to eat.
   Food sources are an important part of habitat.
   Some predators help people.
   Prey have defenses to help protect them from preda-
      tors.

✻ indicates supplementary activity

### Art

Fishing Permits SU-11
Camouflage painting SU-15 ✄

### Blocks/Pretend Play

Let's Go Fishing SU-12
Supplies to cook fish SU-12 ✄
Frogging SU-12 ✄

### Field Trips

Wildlife check station F-19 ✄
Aquatic Life SU-9

### Group

Food chains links SU-14 ✄

### Large Motor/Outside

Play "catch a turkey" F-19 ✄
Compost pile SP-6 ✄
Spider Web Toss SU-2
Play "fish and worms" SU-12 ✄
Foxy Predators SU-14
You Can't Find Me! SU-15

### Manipulative

Predator and prey match SU-14 ✄
Sort animal pictures by skin covering SU-15 ✄

### Music

I'm a Little Fox Squirrel F-18
Who Am I? SP-18
"Ribbit Ribbit" SU-13 ✄

### Nutrition

Animal Harvest F-15
Peanut Butter Playdough Insects SU-2 ✄
Harvest Time SU-16

### Science

Worth Their Weight F-17
Explore a Log SP-7

### Stories/Writing

See How the Turkey Grows F-19
Willie the Woodchuck W-9
Joe's Choice SU-10
Frog and Toads SU-13

# Harvest

Concepts children might construct:
    People harvest plants and animals.
    There are many ways to store and preserve harvest.
    People have many purposes for harvested products.
    Farmers harvest plants and animals to earn money.

### Art

Fishing Permits SU-11

### Blocks/Pretend Play

Toy farm equipment F-12 ✄
Farmer dress-up props F-12 ✄
Produce stand props F-12 ✄
Empty food containers F-13 ✄
Let's go Fishing SU-12
Frogging SU-12 ✄

### Bulletin Board

Farm equipment display F-12 ✄
Seed chart F-12 ✄

### Field Trips

Farm F-12 ✄
Farm implement company F-12 ✄
Grocery store F-12 ✄
Farmer's market F-12 ✄
Apple orchard F-14 ✄
Wildlife check station F-19 ✄
Poultry farm F-20 ✄

### Group

Discuss where food was grown and harvested F-15 ✄

### Large Motor/Outside

Look for farmers harvesting crops F-12 ✄
Pumpkin beanbag toss F-12 ✄
Play "worm through an apple" F-14 ✄

### Manipulative

Fruit/vegetable memory/sorting F-12 ✄
Grind corn F-13 ✄
Weigh, measure and examine apples F-14 ✄
Apple peeler F-14 ✄
Plant and animal product puzzles F-15 ✄
Grind wheat SU-16 ✄

---

✄ indicates supplementary activity

### Nutrition

Vegetable soup F-12 ✸
Popcorn F-13 ✸
Johnny cakes F-13 ✸
Apple Pizzas F-14
Animal Harvest F-15
Make butter F-15 ✸
Harvest Time SU-16

### Science

Something Corny F-13
Apple tasting F-14 ✸
Oxidation of an apple experiment F-14 ✸
Examine wool, feathers, honeycomb, etc. F-15 ✸

### Stories/Writing

The Harvest F-12
Make food books F-12 ✸
See How the Turkey Grows F-19
B, T, and U W-13
Joe's Choice SU-10
*The Little Red Hen* SU-16 ✸

### Woodworking

Woodworking Comparisons F-7

# Insects and Spiders

Concepts children might construct:
    Insects and spiders are similar but different.
    Insects have six legs, three body parts, and wings.
    Spiders have eight legs, two body parts, and build
      webs.
    Insects and spiders can be helpful and harmful to
      people.
    Spiders can be classified by their web.
    Insects and spiders have certain habitat needs.
    Not all crawly creatures are insects or spiders.

### Art

Cave life F-16
Insect creations SP-16 ✸
Folded paper butterfly/moth painting SP-17 ✸
Butterfly and moth sketches SP-17 ✸
Butterfly and moth creations SP-17 ✸
What Hatches from an Egg? SP-19
Create spiders SU-1 ✸
Spider web drawings SU-2 ✸

Sketch roly-polies SU-5 ✸
Camouflage painting SU-15 ✸

### Blocks/Pretend Play

Butterfly/moth dress-ups SP-17 ✸

### Bulletin Board

Insect pictures SP-16 ✸

### Field Trips

Rotting stump SP-16 ✸
Spider sniffing SU-1 ✸
Spider web painting SU-2 ✸
Capture fireflies SU-3 ✸
Aquatic Life SU-9

### Group

Invite an entomologist SU-4 ✸

### Large Motor/Outside

What Lives in the Soil? SP-10
Listen for insect sounds SP-16 ✸
Look for butterflies and moths SP-17 ✸
Pantomime butterfly/moth metamorphosis SP-17 ✸
Butterfly nets SP-17 ✸
Spider Web Toss SU-2
Adopt a spider SU-2 ✸
Fool fireflies SU-3 ✸
Ant Café SU-4
Insect scavenger hunt SU-4 ✸
You Can't Find Me! SU-15

### Manipulative

Where Do Animals Go When It Rains? SP-3
Sort animal pictures by those that hatch from eggs
    and those that don't SP-19 ✸
Sort animal pictures by number of legs SU-1 ✸
Spider lotto SU-2 ✸

### Music

My Friend Little Caterpillar SP-17
The Lights Go On SU-3
"Ribbit Ribbit" SU-13 ✸

### Nutrition

Marshmallow and pretzel spiders SU-1 ✸
Peanut Butter Playdough Insects SU-2 ✸

✸ indicates supplementary activity

## Science

Worth Their Weight F-17
What's an insect? SP-16
Explore a Log SP-7
Observe and hop like a grasshopper or cricket SP-16 ✄
Observe caterpillars SP-17 ✄
What Has Eight Legs? SU-1
Spider collection SU-2 ✄
Place jar of fireflies in cold water SU-3 ✄
Roly-Poly Paradise SU-5
Roly-poly experiments SU-5 ✄

## Stories/Writing

See How the Turkey Grows F-19
Class butterfly book SP-17 ✄
Frogs and Toads SU-13

# Land use

Concepts children might construct:
    People choose to use land in many different ways.
    Land is a natural resource.
    People need to carefully consider how they use land.
    People and animals are dependent on land.

## Art

Winter activity collages W-10 ✄
Snow sculptures W-10 ✄
Noise makers SP-14 ✄
Fishing Permits SU-11

## Blocks/Pretend Play

Community helper props F-1 ✄
Toy farm equipment F-12 ✄
Farmer dress-up props F-12 ✄
Produce stand F-12 ✄
Ice fishing W-10 ✄
Snowmobiles W-10 ✄
Carpet sleds W-10 ✄
Flower Shop SP-12
Let's Go Camping! SP-15
Garbage collector props SU-6 ✄
Let's Go Fishing SU-12
Frogging SU-12 ✄
Mining dress-ups SU-18 ✄

## Bulletin Board

Pictures of children participating in various outdoor
    seasonal sports W-10 ✄

Litter scene SU-6 ✄
Concrete Workers Dress Ups SU-19 ✄

## Field Trips

What Can You See? F-1
Various types of land use areas F-1 ✄
Visit same land use area at different times of the year
    F-1 ✄
Apple orchard F-14 ✄
Power plant W-12 ✄
Service station W-12 ✄
Trash Pickup SU-6
Landfill/recycling plant SU-6 ✄

## Large Motor/Outside

Campfire props F-3 ✄
Look for farmers harvesting crops F-12 ✄
Play catch a turkey F-20 ✄
Skaters Away W-10
Snow shovels W-10 ✄
Winter Olympics W-10 ✄
Compost pile SP-6 ✄
Boat in play yard SU-12 ✄
Practice casting SU-12 ✄
Concrete Hand Impressions SU-19

## Music

"This is the way we shovel the snow" W-10 ✄

## Nutrition

Vegetable soup F-12 ✄
Animal Harvest F-15
Harvest Time SU-16

## Science

Hang On! F-4
Something Corny F-13
Birdfeeders W-1
How Does Your Garden Grow? SP-11
What Is a Mineral SU-18

## Stories/Writing

Molly and the Forest Fire F-3
The Harvest F-12
Food books F-12 ✄
See How the Turkey Grows F-19
B, T, and U W-13
Fred's Forest W-18
Joe's Choice SU-10

✄ indicates supplementary activity

## Woodworking

Woodworking Comparisons F-7

# Mammals

Concepts children might construct:
> Mammals are warm-blooded.
> Mammals nurse their young.
> Mammals have live births.
> Mammals have hair or fur on their bodies.
> Mammals have a keen sense of smell.
> Mammals have certain habitat requirements.
> Some mammals are predator or prey.
> Mammals can be identified by their tracks.

## Art

Cave Life F-16
Sponge paint bats F-17 ✄
Bare feet prints W-8 ✄
Hunting permits SU-10 ✄
Camouflage painting SU-15 ✄

## Blocks/Pretend Play

Shelters for animals to hibernate W-9 ✄
Baby animals and their mothers SP-20 ✄
Materials for beaver dams SU-9 ✄
Conservation agent dress-up clothes SU-10 ✄

## Bulletin Board

Shoe track rubbings W-8 ✄
Baby animals and their mothers SP-20 ✄

## Field Trips

Look for signs of squirrels F-18 ✄
Look for tracks W-8 ✄
Farm with baby animals SP-20 ✄

## Group

Echo experiments F-17 ✄
Do You Smell My Mother? SP-20

## Large Motor/Outside

Blind-folded obstacle course F-17 ✄
Peanut butter cookie sheets W-8 ✄
Sand/snow tracks W-8 ✄
Look for animal signs W-8 ✄
Look for places to hibernate W-9 ✄

Find warm and cold places on play yard W-9 ✄
Look for dormant plants on play yard W-9 ✄
Bunny hop races/walk SP-18 ✄
Foxy Predators SU-14
You Can't Find Me! SU-15

## Manipulative

Track Puzzlers W-8
Where Do Animals Go When It Rains? SP-3
Baby and mother animal match SP-20 ✄
Match pictures of predators with prey SU-14 ✄
Sort animal pictures by skin covering SU-15 ✄

## Music

"Hairy not scary" F-17 ✄
I'm a Little Fox Squirrel F-18
Who Am I? SP-18

## Nutrition

Bat shaped fruit leather F-17 ✄
Eat in preparation for hibernation W-9 ✄
Animal crackers and water SP-3 ✄
Rabbit food SP-18 ✄
Smell snack SP-20 ✄

## Science

Worth Their Weight F-17
Track identification guide W-8 ✄
Smell match SP-20 ✄

## Stories/Writing

Molly and the Forest Fire F-3
See How the Turkey Grows F-19
Willie the Woodchuck W-9
Seeds, Roots, and Plants SP-14

# Plants and flowers

Concepts children might construct:
> Plants are an essential part of the food chain.
> Plants often need insects and animals.
> Plants have habitat requirements.
> Plants reproduce with flowers and seeds.
> Plants can be identified by their flower.
> Flowers have a purpose for the plant's survival.
> Plants and flowers have many identifiable parts.

✄ indicates supplementary activity

### Art

Milkweed seed collages F-11 ✤
Flower collages SP-12 ✤
Flower rubbings SP-13 ✤

### Blocks/Pretend Play

Farm equipment SP-11 ✤
Gardening props SP-11 ✤
Flower Shop SP-12
Harvest flowers SP-12 ✤

### Bulletin Board

Velcro garden SP-11 ✤

### Field Trips

Look for mosses SP-11 ✤
Wildflower habitat SP-12 ✤
Aquatic Life SU-9

### Group

What Is Erosion? SP-9
Dandelion Potpourri SP-13

### Large Motor/Outside

Nature Jar F-10
Look for dormant plants in winter W-9 ✤
Find warm and cold places on play yard W-9 ✤
Class dandelion SP-13 ✤

### Manipulative

Flower match SP-12 ✤
Seeds and plant match SP-14 ✤
Plant puzzle SP-14 ✤

### Music

I'm a Little Milkweed Cradle F-11

### Nutrition

Harvest Time SU-16

### Science

Hang On! F-4
Do Trees Get Drinks? F-5
Plant milkweed seeds F-11 ✤
Place plants in sun and dark W-13 ✤
How Does Your Garden Grow? SP-11
Plant seeds in various soils SP-11 ✤
Water plants with various liquids SP-11 ✤

Force bulbs SP-14 ✤
Identify plant parts SP-14 ✤
Color Queen Anne's lace SP-14 ✤

### Stories/Writing

The Harvest F-12
B, T, and U W-13
Seeds, Roots, and Plants SP-14

### Woodworking

Flower press SP-13 ✤

# Recycling and pollution

Concepts children might construct:

Caring for our natural resources is important.
People can recycle to prevent waste.
Pollution is ugly and harmful to people, plants, and
   animals.
Conservation means "wise use" of our natural
   resources.

### Art

Energy collages W-13 ✤
Sun pictures W-14 ✤
Discuss original purpose of recycled art materials W-17
   ✤
Litterbags SU-6 ✤

### Blocks/Pretend Play

Refueling station for cars and trucks W-13 ✤
Grocery Shopping W-17
Dump trucks and cranes to haul trash SU-6 ✤
Sort trash SU-6 ✤
Garbage collection props SU-6 ✤
Oil and water clean-up SU-8 ✤

### Bulletin Board

Documentation from where do all the dead leaves go?
   SP-6 ✤
Outdoor scene with litter SU-6 ✤

### Field Trips

Wood stove store W-15 ✤
Places that provide energy W-17 ✤
Signs of erosion SP-9 ✤
Trash Pickup SU-6
Landfill/recycling plant SU-6 ✤
Gather and examine roadside leaves SU-8 ✤

### Group

What Is Energy? W-12
No Electricity W-15
Saving Energy W-16
Where Do All the Dead Leaves Go? SP-6
What Is Erosion? SP-9

### Large Motor/Outside

Refueling station for tricycles W-13 ✄
Compost pile SP-6 ✄

### Nutrition

Sun tea W-13 ✄
Sun dried fruit W-14 ✄

### Science

Packing peanuts experiment SP-6 ✄
What's in the Air? SU-8

### Stories/Writing

B, T, and U W-13
Fred's Forest W-18

### Woodworking

Dismantle appliances W-12 ✄

# Rocks

Concepts children might construct:
    People use rocks for many things.
    There are many kinds of rocks.
    People and nature can make rocks shiny.
    Rocks are an important natural resource.

### Art

Cave Life F-16
Granite paper SU-19 ✄
Watercolor rocks SU-20 ✄

### Blocks/Pretend Play

Rock hound props SU-17 ✄
Mining clothes SU-18 ✄
Pea gravel in sensory table SU-19 ✄
Concrete worker props SU-19 ✄

### Bulletin Board

Documentation of crystal garden SU-18 ✄
Documentation of quest for shiny rocks SU-20 ✄

### Field Trips

Cave F-16 ✄
Dig clay SP-8 ✄
Rock Collection SU-17
Art/anthropology museum SU-17 ✄
Rock quarry SU-19 ✄
Pouring concrete SU-19 ✄

### Group

Objects made from minerals SU-18 ✄

### Large Motor/Outside

Outside kiln for clay firing SP-8 ✄
Look for rocks on play yard SU-17 ✄
Concrete Hand Impressions SU-19

### Manipulative

Sort rocks by color, size, texture, etc. SU-17 ✄
Rock memory SU-17 ✄
Gemstone match SU-17 ✄

### Science

Weigh rocks SU-17 ✄
Rock texture feely box SU-17 ✄
What Is a Mineral? SU-18
Rock display SU-19 ✄
Quest for shiny rocks SU-20
Rock tumbler SU-20 ✄

# Seeds

Concepts children might construct:
    Seeds are the way plants reproduce.
    Seeds have many ways of traveling.
    Seeds are a food source for many animals and people.
    Seeds grow into plants when planted.
    Most plants produce many seeds.

### Art

Seed collages F-10 ✄
Milkweed pod collages F-11 ✄

### Blocks/Pretend Play

Farm equipment F-12 ✄
Gardening props SP-11 ✄

### Bulletin Board

Garden planting SP-11 ✄

✄ indicates supplementary activity

### Group

Play "what's missing?" with seeds F-10 ✄
Discuss where animals bury food F-18 ✄

### Large Motor/Outside

Nature Jar F-10
Collect seeds and berries for birds F-10 ✄
Blow seeds F-11 ✄
Look for seeds that travel in wind F-11 ✄
Look for farmers harvesting crops F-12 ✄
Blow on dandelion gone to seed SP-13 ✄
Class dandelion SP-13 ✄

### Manipulative

Grind corn F-13 ✄
Count seeds and weigh apples F-14 ✄
Grind wheat SU-18 ✄

### Music

I'm a Little Milkweed Cradle F-11
I'm a Little Fox Squirrel F-18
I'm a Very Fine Turkey F-20

### Nutrition

Seed of the day snack F-10 ✄
Popcorn F-13 ✄
Harvest Time SU-16

### Science

Examine milkweed pods F-11 ✄
Plant milkweed seeds F-11 ✄
Something Corny F-13
Birdfeeders W-1
Plant birdseed W-1 ✄
How Does Your Garden Grow? SP-11

### Stories/Writing

The Harvest F-12
Make food books F-12 ✄
Seeds, Roots, and Plants SP-14

# Soil

Concepts children might construct:
    People use soil for many purposes.
    Plants need soil to grow.
    Plants keep soil in place.
    Soil is made up of ground-up rocks, and plant and
      animal matter.

Many creatures live in and help soil.
There are different types of soil
Soil will erode if we don't take care of it.

### Art

Mud paint SP-8 ✄
Soil colored paint SP-8 ✄
Soil creatures SP-10 ✄

### Blocks/Pretend Play

Toy farm equipment F-12 ✄
Farmer dress-up props F-12 ✄
Soil and worms in sensory table SP-10 ✄
Pretend to be soil animals SP-10 ✄
Gardening props SP-11 ✄

### Bulletin Board

Documentation from where do all the dead leaves go?
   SP-6 ✄
Documentation from what is erosion? SP-9 ✄

### Field Trips

Visit site with exposed tree roots F-4 ✄
Dig clay SP-8 ✄
Look for erosion SP-9 ✄

### Group

Where Do All the Dead Leaves Go? SP-6
What Is Erosion? SP-9

### Large Motor/Outside

Compost pile SP-6 ✄
Mud Pies SP-8
What Lives in the Soil? SP-10

### Science

Hang On! F-4
Explore a Log SP-7
Bring soil from home SP-8 ✄
Mudpies from various soils SP-8 ✄
Worm ranch SP-10 ✄
Earthworm tracks SP-10 ✄
How Does Your Garden Grow? SP-11
Plant seeds in various soils SP-11 ✄

### Stories/Writing

The Harvest F-12
Class mud pie recipe book SP-8 ✄
Seeds, Roots, and Plants SP-14

---

✄ indicates supplementary activity

# Trees and Forests

Concepts children might construct:

Trees are beneficial to people and animals in many
ways (provide shade, habitat, clean the air, hold soil
in place, food, etc.).

People sometimes harvest trees.

Trees are a renewable resource.

Trees can be identified by leaf, bark, seeds, nuts, etc.

Trees have several parts.

Trees can be destroyed by fire.

## Art

Forest fire pictures F-3 ✀
Tree pictures F-6 ✀
Leaf collages F-9 ✀
Leaf prints and rubbings F-9 ✀

## Blocks/Pretend Play

Firefighter dress-ups F-3 ✀
Tree surgeon or carpenter dress-up props F-7 ✀
Fruits and nuts F-7 ✀
Twig trees F-8 ✀

## Bulletin Board

Match tree rubbings with leaves F-3 ✀
Forest Fire Pictures F-3 ✀
Documentation of hang on! F-4 ✀
Leaf rubbings match F-10 ✀

## Field Trips

Recreational site and build a campfire F-3 ✀
Look for exposed tree roots F-4 ✀
Lumber yard, furniture shop, sawmill, or instrument
maker F-7 ✀
Apple orchard F-14 ✀
Wood stove store W-12 ✀
Forest W-17 ✀
Rotting log SP-7 ✀

## Group

Invite a fiddler F-7 ✀
Invite a woodworker F-7 ✀
Find leaf partners F-9 ✀
Discuss forest products W-18 ✀
Where Do All the Dead Leaves Go? SP-6
Explore a Log SP-7

## Large Motor/Outside

Tree Skin F-2

Leaf scavenger hunt F-2 ✀
Look for surface tree roots F-4 ✀
Adopt a tree F-8 ✀
Look for wildlife using trees F-8 ✀
Wet leaf prints F-9 ✀
Match leaf colors F-9 ✀
Nature Jar F-10

## Manipulative

Tree Puzzle F-6
Sort animal pictures by those that live in trees and
those that don't F-8 ✀
Leaf Lotto F-9
Leaf memory or concentration F-9 ✀
Sort leaves F-9 ✀
Weigh and measure apples F-14 ✀
Apple peeler F-14 ✀
Where Do Animals Go When It Rains? SP-3

## Music

"What kind of tree are you?" F-6 ✀
I'm a Little Fox Squirrel F-18
I'm a Very Fine Turkey F-20

## Nutrition

Fruits or nuts harvested from local trees F-7 ✀
Apple Pizzas F-14

## Science

Examine tree "cookies" F-2 ✀
Examine charred wood F-3 ✀
Hang On! F-4
Do Trees Get Drinks? F-5
Soak leaves in water F-5 ✀
Apple tasting F-14 ✀
Birdfeeders W-1
Explore a Log SP-7

## Stories/Writing

Molly and the Forest Fire F-3
Tree Books F-8
Individual tree books F-8 ✀
Johnny Appleseed F-14 ✀
See How the Turkey Grows F-19
Fred's Forest W-18

## Woodworking

Woodworking Comparisons F-7

✀ indicates supplementary activity

# Weather

Concepts children might construct:
> Weather changes with each season.
> Weather forces people and animals to change and adapt.
> Seasons are predictable.
> Shadows are related to the sun or other light source.

## Art

Winter activity collages W-10 ✀
Snow sculptures W-10 ✀
Snow shakers W-10 ✀
Ice sculptures W-10 ✀
Clothing collages W-11 ✀
Fabric collages W-20 ✀
Rainbows SP-1 ✀
Cloud pictures SP-1 ✀
Storm finger paint SP-1 ✀
What Goes in the Wind? SP-4
Wind chimes SP-4 ✀
Blow painting SP-4 ✀
Add shadows to drawings SP-5 ✀

## Blocks/Pretend Play

Hibernation cave W-9 ✀
Ice fishing W-10 ✀
Snowmobiles W-10 ✀
Carpet sleds W-10 ✀
Weather station W-11 ✀
Seasonal clothing W-11 ✀
Rain gear SP-1 ✀
Hanging Out SP-2
Add shelters for animals in rain SP-3 ✀

## Bulletin Board

Display pictures of children participating in various outdoor seasonal sports W-10 ✀
Display pictures of winter scenes or animals in winter environments W-11 ✀
Display pictures of people in seasonal clothing W-20 ✀
Cloud photographs and children's ideas about them SP-1 ✀
Display pictures of how wind helps and harms the environment SP-4 ✀
Children and their shadows SP-5 ✀

## Field Trips

Walk in the rain SP-1 ✀

## Group

Discuss where animals bury their food F-18 ✀
Chart outdoor sports preferences W-10 ✀
Chart temperature and weather W-11 ✀
Record and chart number of children wearing weather related clothing W-11 ✀
Play "whose shadow?" SP-5 ✀

## Large Motor/Outside

Skaters Away W-10
Freeze tag W-10 ✀
Winter Olympics W-10 ✀
Shovel snow W-10 ✀
Notice weather related animal activity patterns W-11 ✀
Spray-paint snow W-14 ✀
Rainbow dances SP-1 ✀
Observe clouds SP-1 ✀
Pretend to be raindrops SP-1 ✀
Rain gauge SP-1 ✀
Water paint SP-2 ✀
Crepe paper streamers SP-4 ✀
Kites SP-4 ✀
Bubbles SP-4 ✀
Shadow Hunt SP-5
Shadow dance in projector light SP-5 ✀
Shadow tag SP-5 ✀

## Manipulative

Tree Puzzle F-6
Sort pictures of clothing by season W-11 ✀
Where Do Animals Go When It Rains? SP-3
Shadow and animal match SP-5 ✀

## Music

I'm a Little Fox Squirrel F-18
"This is the way we shovel the snow" W-10 ✀
Rain rhythm SP-1 ✀
Three rain drops SP-1 ✀
Rain stick SP-1 ✀
Rain recordings SP-1 ✀

## Nutrition

Marshmallow snow sculptures W-10 ✀
Mashed potato clouds SP-1 ✀
Animal crackers and water SP-3 ✀

## Science

What's the Temperature? W-11
Insulation W-19

---

✀ indicates supplementary activity

Winter fabrics W-20 ✀
Prisms SP-1 ✀
Cloud drawings SP-1 ✀
Rainfall chart SP-1 ✀
Observe a puddle SP-2 ✀
Add color to shadows SP-5 ✀
Draw shadows at various times of day SP-5 ✀

## Stories/Writing

See How the Turkey Grows F-19
Willie the Woodchuck W-9
B, T, and U W-13
Fred's Forest W-18
Drip and Drop SP-1
Seeds, Roots, and Plants SP-14

✀ indicates supplementary activity

# Appendix D

## Patterns

**Fall Pattern 1**

**Fall Pattern 2**

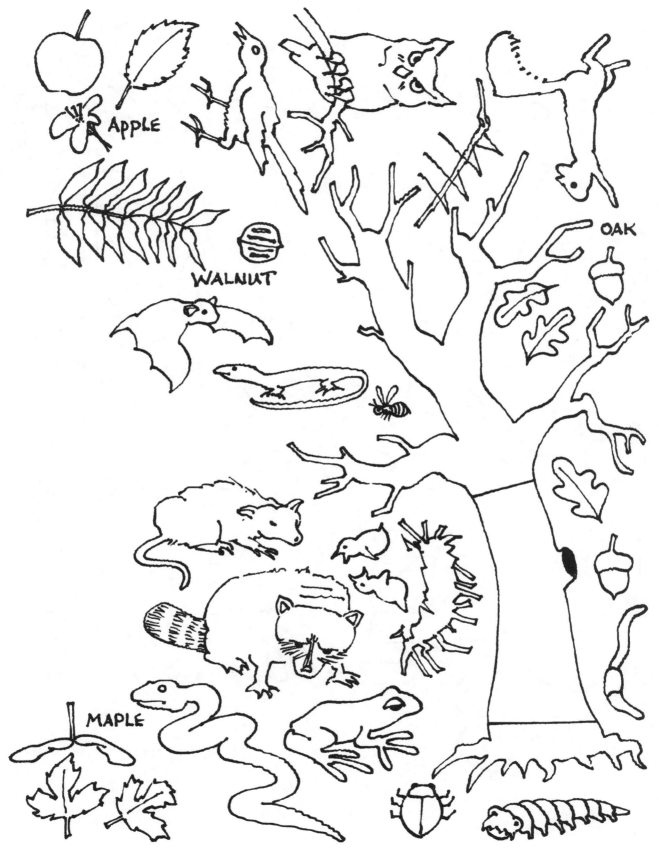

APPLE

WALNUT

OAK

MAPLE

**Fall Pattern 3**

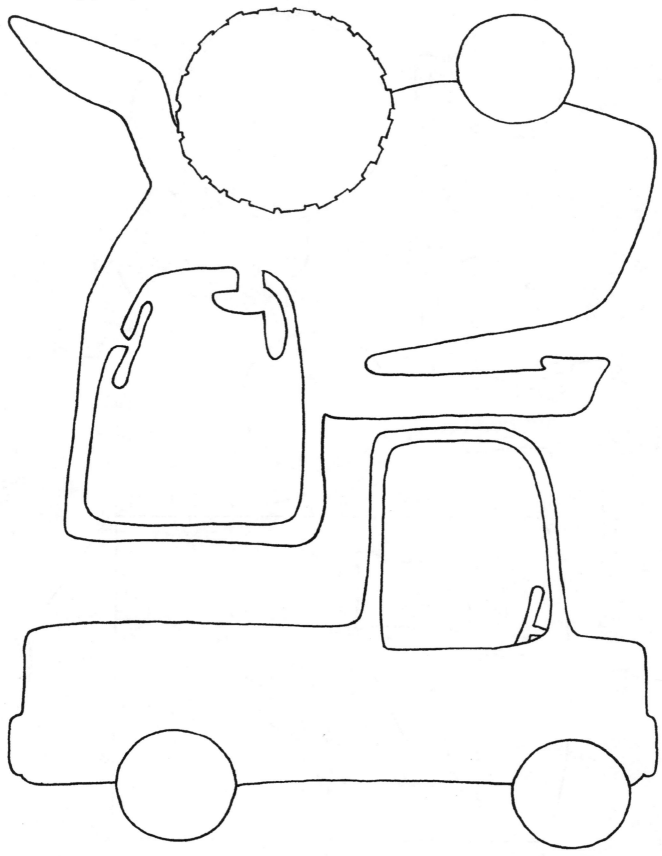

**Fall Pattern 4**

# Winter Pattern 1

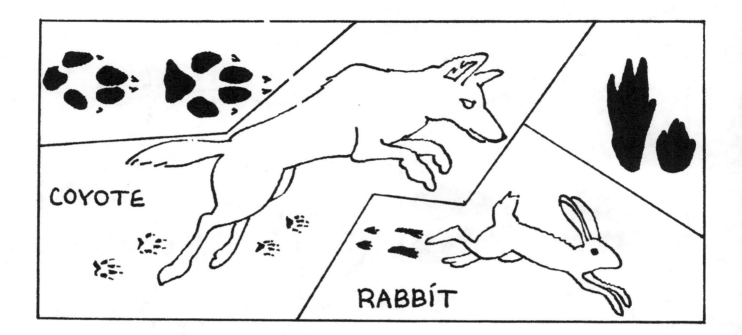

DEER

**Winter Pattern 3**

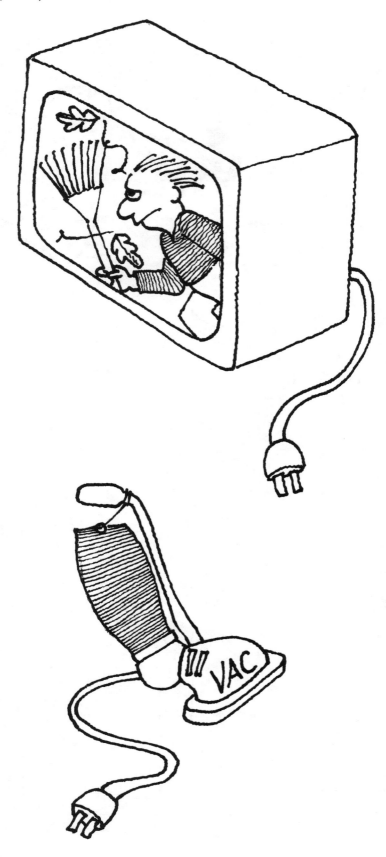

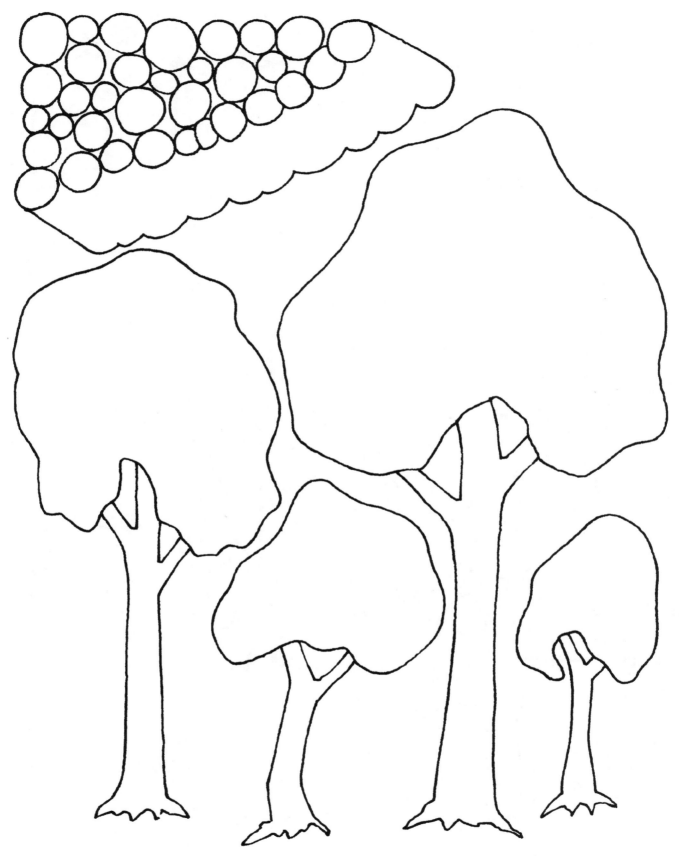

**Spring Pattern 1**

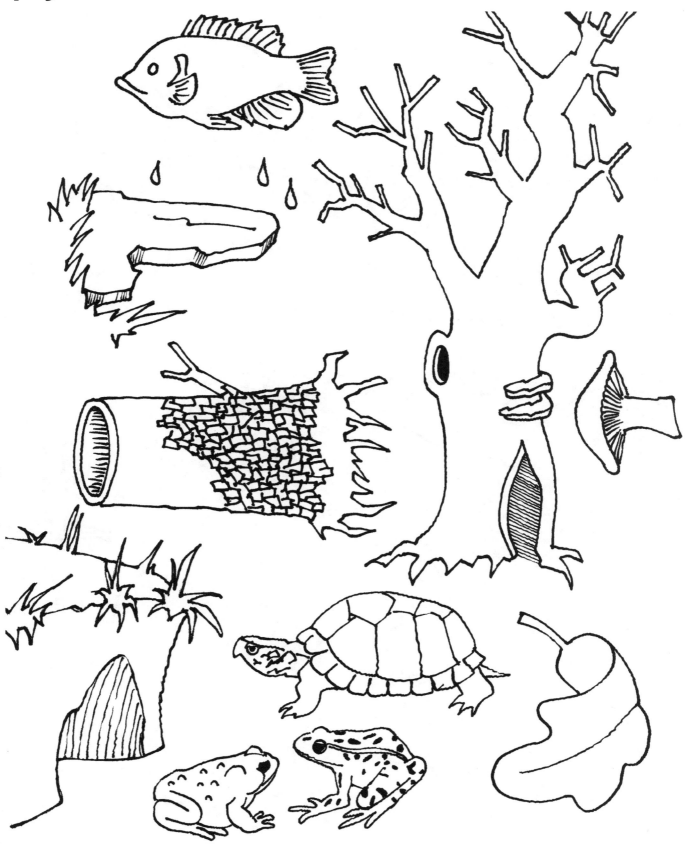

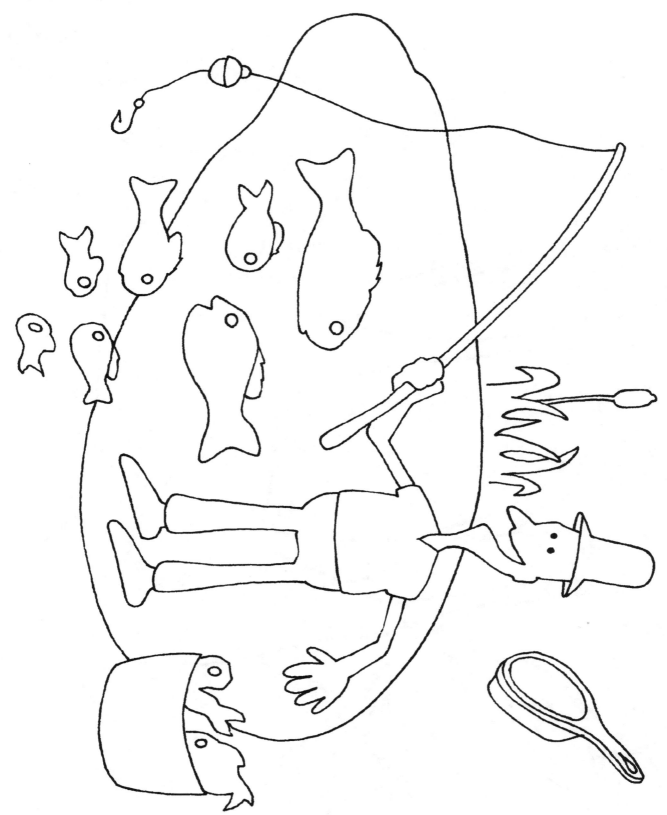

# Summer Pattern 2

# Other Resources from Redleaf Press

**HOLLYHOCKS AND HONEYBEES: GARDEN PROJECTS FOR YOUNG CHILDREN**
*by Sara Starbuck, Marla Olthof, and Karen Midden*
This practical guide introduces teachers—with or without green thumbs—to the rich learning opportunities found in gardening with children.

**THEME KITS MADE EASY**
*by Leslie Silk Eslinger*
*Theme Kits Made Easy* contains a tested method for putting together creative kits using dozens of theme suggestions. Create your own kits using theme ideas like Farm, Naptime, and Three Billy Goats Gruff. Includes suggestions for choosing anti-bias materials and resources.

**LEARN AND PLAY THE RECYCLE WAY: HOMEMADE TOYS THAT TEACH**
*by Rhoda Redleaf and Audrey S. Robertson*
Features the best activities from *Teachables from Trashables* and new ways to make learning fun with homemade toys. Children will love making drums from coffee cans, cactus gardens from jars, and planters from plastic bottles. Sections for infants, toddlers, preschoolers, and schoolagers.

**LESSONS FROM TURTLE ISLAND: NATIVE CURRICULUM IN EARLY CHILDHOOD CLASSROOMS**
*by Guy W. Jones and Sally Moomaw*
The first complete guide to exploring Native American issues with children. Includes five cross-cultural themes—Children, Home, Families, Community, and the Environment.

**MORE THAN MAGNETS: EXPLORING THE WONDERS OF SCIENCE IN PRESCHOOL AND KINDERGARTEN**
*by Sally Moomaw and Brenda Hieronymus*
*More Than Magnets* takes the uncertainty out of teaching science. More than 100 activities engage children in interactive science opportunities—including life science, physics, and chemistry activities.

**800-423-8309**

**www.redleafpress.org**

## DATE DUE

DEC 1 6 2013

GAYLORD

PRINTED IN U.S.A.